Routledge Guides to the Great Books

The Routledge Guidebook to Heidegger's *Being and Time*

Being and Time is considered to be one of the most important philosophical texts of the twentieth century and its influence can be seen in existentialism, metaphysics, postmodernism and more. But the author, Martin Heidegger, was a controversial figure and his writing can be very daunting for the first time reader.

The Routledge Guidebook to Heidegger's Being and Time offers a route through this classic text by exploring:

- The context of Heidegger's work and the background to his writing
- Each separate part of the text in relation to its goals, meanings and impact
- The enduring legacy and influence of Heidegger's work

Following Heidegger's text closely, this guidebook is essential reading for all students of philosophy, and all those wishing to gain an understanding of this classic work.

Stephen Mulhall *is Fellow & Tutor in Philosophy at New College, University of Oxford.*

ROUTLEDGE GUIDES TO THE GREAT BOOKS

Series Editor: Anthony Gottlieb

The Routledge Guides to the Great Books provide ideal introductions to the work of the most brilliant thinkers of all time, from Aristotle to Marx and Newton to Wollstonecraft. At the core of each Guidebook is a detailed examination of the central ideas and arguments expounded in the great book. This is bookended by an opening discussion of the context within which the work was written and a closing look at the lasting significance of the text. *The Routledge Guides to the Great Books* therefore provide students everywhere with complete introductions to the most important, influential and innovative books of all time.

Available:

Aristotle's Nicomachean Ethics Gerard J. Hughes
Hegel's Phenomenology of Spirit Robert Stern
Heidegger's Being and Time Stephen Mulhall
Locke's Essay Concerning Human Understanding E. J. Lowe
Plato's Republic Nickolas Pappas
Wollstonecraft's A Vindication of the Rights of Woman Sandrine Bergès
Wittgenstein's Philosophical Investigations Marie McGinn

Forthcoming:

De Beauvoir's The Second Sex Nancy Bauer
Descartes' Meditations on First Philosophy Gary Hatfield
Galileo's Dialogue Maurice A. Finocchiaro
Hobbes' Leviathan Glen Newey
Mill's On Liberty Jonathan Riley

Routledge Guides to the Great Books

The Routledge Guidebook to:
Heidegger's *Being and Time*

Stephen Mulhall

Routledge
Taylor & Francis Group
LONDON AND NEW YORK

First edition published in the Routledge Philosophy Guidebook Series in 1996
Second edition published in the Routledge Philosophy Guidebook Series in 2005
First published in The Routledge Guides to the Great Books Series in 2013
by Routledge
2 Park Square, Milton Park, Abingdon, Oxon OX14 4RN

Simultaneously published in the USA and Canada
by Routledge
711 Third Avenue, New York, NY 10017

Routledge is an imprint of the Taylor & Francis Group, an informa business

British Library Cataloguing in Publication Data
A catalogue record for this book is available from the British Library

Library of Congress Cataloging in Publication Data
The Routledge guidebook to Heidegger's Being and time / Stephen Mulhall.
     p. cm. – (The Routledge guides to the great books)
   Rev. ed. of: Routledge philosophy guidebook to Heidegger and Being and time.
   Includes bibliographical references (p.) and index.
   1. Heidegger, Martin, 1889-1976. Sein und Zeit. I. Mulhall, Stephen, 1962- Routledge philosophy guidebook to Heidegger and Being and time. II. Title.
   B3279.H48S46654 2013
   111–dc23
                2012016463

ISBN: 978-0-415-66442-4 (hbk)
ISBN: 978-0-415-66444-8 (pbk)
ISBN: 978-0-203-08431-1 (ebk)

Typeset in Aldus and Scala by
Florence Production Ltd, Stoodleigh, Devon

CONTENTS

PREFACE

Martin Heidegger was born in Messkirch on 26 September 1889. An interest in the priesthood led him to commence theological and philosophical studies at the University of Freiburg in 1909. A monograph on the philosophy of Duns Scotus brought him a university teaching qualification, and in 1922 he was appointed to teach philosophy at the University of Marburg. The publication of his first major work, *Sein und Zeit (Being and Time)*, in 1927 catapulted him to prominence and led to his being appointed to the Chair of Philosophy at Freiburg in 1928, succeeding his teacher and master, the phenomenologist Edmund Husserl. From April 1933 until his resignation in February 1934, the early months of the Nazi regime, he was Rector of Freiburg. His academic career was further disrupted by the Second World War and its aftermath: in 1944, he was enrolled in a work-brigade, and between 1945 and 1951 he was prohibited from teaching under the deNazification rules of the Allied authorities. He was reappointed Professor in 1951, and gave occasional seminars in his capacity as Honorary Professor until 1967, as well as travelling widely and participating in conferences and colloquia on his work. He continued to write until his death on 26 May 1976. He is buried in the local grave-yard of his birthplace, Messkirch.

This brief biographical sketch leaves much that is of importance in Heidegger's life (particularly his destructive and ugly relations with Nazism) unexplored; but it gives even less indication of the breadth, intensity and distinctiveness of his philosophical work and its impact on the development of the discipline in Europe. The publication of *Being and Time* transformed him from a charismatic lecturer, well known in German academic life (Hannah Arendt said that descriptions of his lecture series circulated in Germany as if they were 'rumours of a hidden king'), into a figure of international significance. A steady stream of lectures, seminars and publications in the following decades merely broadened and intensified his influence. Sartrean existentialism, the hermeneutic theory and practice of Gadamer, and Derridean decon-struction all grew from the matrix of Heidegger's thought; and the cognate disciplines of literary criticism, theology and psychoanalysis were also importantly influenced by his work. To some, his preoccu-pations – and, more importantly, the manner in which he thought and wrote about them – signified only pretension, mystification and char-latanry. For many others, however, the tortured intensity of his prose, its breadth of reference in the history of philosophy, and its arrogant but exhilarating implication that nothing less than the continuation of Western culture and authentic human life was at stake in his thought, signified instead that philosophy had finally returned to its true con-cerns in a manner that might justify its age-old claim to be the queen of the human sciences.

This book is an introduction for English-speaking readers to the text that publically inaugurated Heidegger's life-long philosophical project – *Being and Time*.[1] It aims to provide a perspicuous surview of the structure of this complex and difficult work, clarifying its under-lying assumptions, elucidating its esoteric terminology and sketching the inner logic of its development. It takes very seriously the idea that it is intended to provide an introduction to a text rather than a thinker or a set of philosophical problems. Although, of course, it is not possible to provide guidance for those working through an extremely challenging philosophical text without attempting to illum-inate the broader themes and issues with which it grapples, as well as the underlying purposes of its author, it is both possible and desirable to address those themes and purposes by relating them very closely and precisely to the ways in which they are allowed to emerge in the chapter by chapter, section by section structure of the text concerned. This introduction is therefore organized in a way that is designed to mirror that of *Being and Time* as closely as is consistent with the demands of clarity and surveyability.

This book is not an introduction to the many important lines of criticism that have been made of Heidegger's book since its first publication. Those criticisms can be properly understood only if one has a proper understanding of their object; and their force and cogency can be properly evaluated only if one has first made the best possible attempt to appreciate the power and coherence of the position they seek to undermine. For these reasons, I have concentrated on providing an interpretation of *Being and Time* which makes the strongest case in its favour, that is consistent both with fidelity to the text and to the canons of rational argument. My concern is to show that there is much that is well worth arguing over in Heidegger's early work; but I do not attempt to judge how those arguments might be conducted or definitively concluded.

As Heidegger himself emphasized, no interpretation of a text can be devoid of preconceptions and value-judgements. Even a basic and primarily exegetical introduction to the main themes of a philosophical work must choose to omit or downplay certain details and complexities, and to organize the material it does treat in one of many possible ways. But my interpretation of *Being and Time* takes up an unorthodox position on a highly controversial issue in Heidegger scholarship; the reader unfamiliar with that scholarship should be warned of this in advance. Particularly with respect to the material in the second half of *Being and Time*, I regard Heidegger's treatment of the question of human authenticity as necessarily and illuminatingly applicable to his conception of his role as a philosopher, and so to his conception of his relation to his readers. In other words, I read his philosophical project not only as analysing the question of what it is for a human being to achieve genuine individuality or selfhood, but as itself designed to facilitate such an achievement in the sphere of philosophy. As will become clear, Heidegger does not conceive of

human authenticity as a matter of living in accord with some particular ethical blueprint; and to this degree, my interpretation cannot
properly be thought of as a moralization of *Being and Time*. It does
imply, however, that the tone of spiritual fervour that many readers
have detected in the book is internally related to its most central
purposes, and that Heidegger makes existential demands on himself
and his readers. This is something that many careful students of *Being
and Time* have been eager to deny. The legitimacy of my interpretative strategy must, of course, ultimately depend upon the conviction
it elicits as a reading of *Being and Time*; but I feel it right to declare
it in advance, and in so doing to declare further that I cannot otherwise make sense of the structure of the book as a whole, and of its
unremitting concern with its own status as a piece of philosophical
writing.

I would like to acknowledge the help various people have given me
in the course of writing this book. My colleagues at the University of
Essex – particularly Simon Critchley and Jay Bernstein – have gener-
ously allowed me to draw upon their extensive knowledge of Heidegger
and Heideggerian scholarship; and Jay Bernstein also commented in
detail on an early draft of my manuscript. The editors of this series
– Tim Crane and Jo Wolff – kindly invited me to take on this project
in the first place, and provided much useful advice as it developed.
Two anonymous readers' reports on the manuscript arrived at a late
stage in its preparation. Both helped to improve the book significantly,
and I would like to thank their authors. Finally, I would also like to
thank Alison Baker for her forbearance and support during my work
on this project.

NOTE

1   All quotations and references are keyed to the standard Macquarrie
    and Robinson translation of the original German text (Oxford: Basil
    Blackwell, 1962). The location of all quotations is given by specifying
    the relevant section and page, in that order e.g. (BT, 59: 336).

PREFACE TO THE
SECOND EDITION

It is now more than a decade since I began work on the first edition of this book. Since then, I have continued to think about Heidegger's philosophical writings in general, and *Being and Time* in particular; and although I continue to believe that the fundamental aspects of my original interpretation of it are sound, I have gradually come to feel that various issues might usefully be explored in more detail or introduced into a discussion that wrongly omitted them.

First, I now realize that my original analysis of Heidegger's treatment of scepticism in Division One of *Being and Time* was importantly incomplete. In the first edition, I concentrated on drawing out his reasons for thinking that a proper understanding of Dasein as Being-in-the-world would render scepticism inarticulable, and thus eliminate what he called the scandalous fact of philosophy's endless and endlessly unsuccessful attempts to refute scepticism, by revealing its essential emptiness. More recently, I have come to believe that this line of argument in *Being and Time* is counterbalanced by a second, more recessive but also more radical one. This depends upon appreciating that scepticism can be understood as having not only a putative

cognitive content or thrust, but also (as with any mode of understanding, according to Heidegger's own analysis) a specific mood or
mode of attunement – that of anxiety or angst. And Heidegger's argument in Division One is that angst is capable of pivoting Dasein from
its lostness in 'das man' to an authentic grasp of itself, the world and
Being. From this, it would seem to follow that philosophical scepticism is inherently capable of disclosing a vital dimension of Dasein's
Being, and so of Being as such, and hence that Heidegger cannot
avoid thinking of scepticism as an essential moment in any philosophical recovery of the question of the meaning of Being.

Second, I have come to see more clearly the peculiar nature, and
the absolutely fundamental importance, of the relation Heidegger
constructs between Divisions One and Two of *Being and Time*. The
argument of Division Two begins from a sense that the analysis of
Division One overlooks an essential aspect of the totality of Dasein's
Being – its relation to its own end. This turns out to involve Dasein's
multiple and determining relationship to its own nothingness, and
hence to negation or nullity more generally; and by the time of his
discussion of Dasein's conscience, it becomes clear that Division Two
intends to draw out the full implications of the relatively glancing
claim in Division One that angst reveals Dasein's Being to be essentially uncanny, or not-at-home in the world. I now think of this as
Dasein's failure or inability to coincide with itself; and this in turn
suggests that what Heidegger means by Dasein's inauthenticity is
its various attempts to live as if it did coincide with itself – as if its
existential potential coincided with its *existentiell* actuality. Hence,
authenticity is a matter of living out Dasein's essential non-identity
with itself; and, accordingly, any authentic analytic of Dasein's Being
must manifest a similar failure of self-identity. Its construction or form
must reflect the fact that any account of Dasein's Being must indicate its own inadequacy, its own ineliminable reference to that which
is beyond Dasein's, and hence its own, grasp.

I would now argue that this is the function of Division Two in
relation to Division One: the former is precisely designed to unsettle
our confidence in the latter, our perhaps unduly complacent sense
that it concludes with a genuinely complete, however provisional,
account of Dasein's Being (in terms of care). In other words, Division

Two does not (or not only) amount to a deeper exploration of the structures established in Division One; it is also an attempt to reveal the ways in which those structures in fact point towards Dasein's essential dependence upon that which exceeds its own limits – and in particular the limits of its own comprehension. One might say that it ensures that *Being and Time* as a whole does not coincide with itself, and thus meets the criterion it establishes for authenticity.

If this view is right, then Division Two cannot be dismissed as concerning itself with more or less marginal matters of ethics and theology – the essentially optional existential side of Heidegger's phenomenology. In particular, the idea that one can give an account of the core of the whole book while limiting oneself to the material of Division One (as Hubert Dreyfus's highly influential commentary, *Being-in the-World*,[1] in effect does) becomes completely untenable. A proper appreciation of that fact alone would radically put in question the ways in which Heidegger's early thought has been appropriated in the Anglo-American philosophical world. It would also illuminate the degree to which the insights of *Being and Time* prefigure the claims Heidegger makes at the beginning of the 1930s (in, for example, his famous inaugural lecture, *What is Metaphysics?*[2]) about an internal relation between Being and 'the nothing' – claims sometimes taken to herald a fundamental turn in his thinking. And, as a result, it would significantly alter our sense of the internal relation of Heidegger's early work to that of Sartre; for if this way of understanding *Being and Time*'s purposes is correct, then a book entitled *Being and Nothingness* might come to seem far less distant from its acknowledged source than is often assumed to be the case.

The publication of this second edition has given me the chance to revise the whole of my commentary in the light of these two main shifts in my thinking about *Being and Time*. This means that Chapters 4, 5 and 8 have been very significantly revised and expanded, and that many matters of fine detail in Chapters 6 and 7 have been slightly but importantly altered to accommodate a very different way of viewing Division Two as a whole. I have also taken the opportunity to correct a number of minor flaws throughout the book – almost always, I believe, matters of style rather than of content. In the end, then, this is a very different text to that of the first edition; but these discontinuities in fact grow rather directly from the main emphases of my initial reading of the text – most obviously, from its insistence that the results of Heidegger's existential analytic of Dasein must necessarily apply to its author and his philosophical activities, and hence will directly inform his conception of the standards against which his own writing must measure itself, and of the transformation it must aim to effect upon its readers. In that sense, I would like to believe that the second edition of this book is essentially a more authentic version of the first.

Stephen Mulhall
New College, Oxford
January, 2005

NOTES

1   Cambridge, Mass.: MIT Press, 1991.
2   In D. F. Krell (ed.), *Basic Writings*, 2nd edn (San Francisco, Calif.: Harper, 1993).

SERIES EDITOR'S PREFACE

"The past is a foreign country," wrote British novelist, L. P. Hartley: "they do things differently there."

The greatest books in the canon of the humanities and sciences can be foreign territory, too. This series of guidebooks is a set of excursions written by expert guides who know how to make such places become more familiar.

All the books covered in this series, however long ago they were written, have much to say to us now, or help to explain the ways in which we have come to think about the world. Each volume is designed not only to describe a set of ideas, and how they developed, but also to evaluate them. This requires what one might call a bifocal approach. To engage fully with an author, one has to pretend that he or she is speaking to us; but to understand a text's meaning, it is often necessary to remember its original audience, too. It is all too easy to mistake the intentions of an old argument by treating it as a contemporary one.

The Routledge Guides to the Great Books are aimed at students in the broadest sense, not only those engaged in formal study. The intended audience of the series is all those who want to understand the books that have had the largest effects.

AJG
October 2012

INTRODUCTION: HEIDEGGER'S PROJECT

(*Being and Time*, §§1–8)

THE QUESTION OF BEING

According to Heidegger, the whole of *Being and Time* is concerned with a single question – the question of the meaning of Being. But what does he mean by the term 'Being'? What, if anything, does it signify? It is no accident that Heidegger provides no clear and simple answer to this question – neither at the opening of his book nor at any later point within it; for, in his view, it will take at least the whole of his book to bring us to the point where we can even ask the question in a coherent and potentially fruitful way. Nevertheless, he also takes a certain, preliminary understanding of Being to be implicit in everything human beings say and do; so it should be possible, even at this early stage, to indicate at least an initial orientation for our thinking.

Late in William Golding's novel *The Spire*,[1] its medieval protagonist – a cathedral dean named Jocelin – has a striking experience as he leaves his quarters:

> Outside the door there was a woodstack among long, rank grass.
> A scent struck him, so that he leaned against the woodstack, care-
> less of his back, and waited while the dissolved grief welled out of
> his eyes. Then there was a movement over his head. . . . He twisted
> his neck and looked up sideways. There was a cloud of angels flashing
> in the sunlight, they were pink and gold and white; and they were
> uttering this sweet scent for joy of the light and the air. They brought
> with them a scatter of clear leaves, and among the leaves a long,
> black springing thing. His head swam with the angels, and suddenly
> he understood there was more to the appletree than one branch. It
> was there beyond the wall, bursting up with cloud and scatter, laying
> hold of the earth and the air, a fountain, a marvel, an appletree.
> . . . Then, where the yard of the deanery came to the river and trees
> lay over the sliding water, he saw all the blue of the sky condensed
> to a winged sapphire, that flashed once.
>
>   He cried out.
>   'Come back!'
> But the bird was gone, an arrow shot once. It will never come back,
> he thought, not if I sat here all day.
>
> (Golding 1964: 204–5)

Jocelin, as if for the first time, is struck by the sheer specificity
of the appletree – its springing branches and trunk, the cloud
and scatter of its leaves and blossom, everything that makes it the
particular thing that it is. He is struck by what one might call
the distinctive mode of its existence or being. The kingfisher, in the
singular sapphire flash of its flight, conveys rather a sense of contingency, of the sheer, transient fact of its existence or being. Together,
then, the appletree and the kingfisher impress upon Jocelin a fused
sense of *how* the world is and *that* the world is; they precipitate an
immeasurable astonishment and wonder at the reality of things, at
the fact of there being a highly differentiated world to wonder
at. It is just such a sense of wonder that Heidegger thinks of as a
response to the Being of things, a response to Being; and he aims
to recover in his readers a capacity to take seriously the question
of its meaning or significance.

For some philosophers, the fact that a passage extracted from a novel can so precisely articulate the ground of Heidegger's questioning might suggest new ways of connecting philosophy, literature and everyday human experience, and of recovering the sense of wonder with which the ancient Greeks held that the true impulse to philosophize originates; but for many others it suggests that to take such questioning seriously is to succumb to adolescent Romanticism. Despite these widespread qualms, however, it is perfectly possible to detect in Heidegger's own introductory remarks a way of providing a more obviously 'legitimate' derivation or genealogy for his question – a more philosophically respectable birth certificate.

In everything that human beings do, they encounter a wide variety of objects, processes, events and other phenomena that go to make up the world around them. Taking a shower, walking the dog, reading a book: all involve engaging with particular things in particular situations, and in ways that presuppose a certain comprehension of their presence and nature. In taking a shower, we show our awareness of the plastic curtain, the shower-head and the dials on the control panel, our understanding of the way in which they relate to one another, and so our grasp of their distinctive potentialities. We cannot walk the dog – choosing the best route, allowing time for shrub-sniffing, shortening the lead at the advent of another dog – without revealing our sense of that creature's nature and its physical expression. Enjoying a thriller on the beach presupposes being able to support its bulk and focus on its pages, to grasp the language in which it is written and the specific constraints and expectations within which novels in that particular genre are written and read.

In short, throughout their lives human beings manifest an implicit capacity for a comprehending interaction with entities as actual and as possessed of a distinctive nature. This capacity finds linguistic expression when we complain that the shower curtain *is* split, or wonder aloud what Fido *is* up to now, or ask where our novel *is*. Since this comprehending interaction seems to be systematically registered by our use of various forms of the verb 'to be', Heidegger describes it as an implicit understanding of what it is for an entity to be, and so as a capacity to comprehend beings as such, to comprehend beings *qua* beings. In other words, it is a capacity to comprehend the Being of beings.

Many of our cultural practices in effect amount to rigorous thematizations of particular forms of this comprehension and its corresponding objects; they constitute modes of human activity in which something that is taken for granted, and so remains undeveloped in other parts of our life, is made the explicit focus of our endeavours. For example, our everyday concern for hygiene may lead us to explore the cleansing properties of water, soap and shampoo, and so to a more general study of the structure of matter. Our life with pets may lead us into a study of domestic species and then of animal life more generally. Our ordinary reading habits may lead us to examine a particular author's style and development, and then to investigate the means by which aesthetic pleasure can be elicited from specific literary genres. In other words, such disciplines as physics and chemistry, biology and literary studies take as their central concern aspects of phenomena that remain implicit in our everyday dealings with them; and the specific theories that are produced as a result go to make up a body of what Heidegger would call *ontic* knowledge – knowledge pertaining to the distinctive nature of particular types of entity.

However, such theory-building itself depends upon taking for granted certain basic ways in which the given discipline demarcates and structures its own area of study; and those foundations tend to remain unthematized by the discipline itself, until it finds itself in a state of crisis. Relativity theory precipitated such a crisis in physics; in biology, similar turmoil was caused by Darwinian theories of natural selection; and, in literary studies, theoretical attacks upon prevailing notions of the author, the text and language have recently performed an analogous function. Such conceptual enquiries are not examples of theories that conform to the standards of the discipline, but rather explore that on the basis of which any such theory could be constructed, the a priori conditions for the possibility of such scientific theorizing. In Heideggerian language, what they reveal are the ontological presuppositions of ontic enquiry.

Here, philosophical enquiry enters the scene. For when physics is brought to question its conception of matter, or biology its conception of life, or literary studies its conception of a text, what is disclosed are the basic articulations of that discipline's very subject matter, that which underlies all the specific objects that the discipline takes as its theme; and that is not, and could not be, within the purview of intra-disciplinary enquiry, because it would be presupposed by any such enquiry. What is needed is a reflection upon those articulations, an attempt to clarify the nature and validity of the most basic conceptualizations of this particular domain; and such a critical clarification is the business of philosophy. In these respects, philosophical enquiry is at once parasitic upon, and more fundamental than, other modes of human enquiry. There could be no philosophy of science without science, and philosophy has no authority to judge the validity of specific scientific theories. But any such theory is constructed and tested in ways that presuppose the validity of certain assumptions about the domain under investigation, assumptions that it can consequently neither justify nor undermine, and which therefore require a very different type of examination. The scientist may well be the best exponent of the practices of inductive reasoning as applied to the realm of nature; but if questions are raised about the precise structure of inductive reasoning and its ultimate justification as a mode of discovering truth, then the abilities of the philosopher come into play.

This is a familiar view of the role of philosophical enquiry in the Western philosophical tradition, particularly since the time of Descartes – at least if we judge by the importance it has assigned to the twin ontological tasks of specifying the essential differences between the various types of entity that human beings encounter, and the essential preconditions of our capacity to comprehend them. To learn about that tradition is to learn, for example, that Descartes' view of material objects – as entities whose essence lies in being extended – was contested by Berkeley's claim that it lies in their being perceived, whereas his view that the essence of the self is grounded in the power of thought was contested by Hume's claim that its only ground is the bundling together of impressions and ideas. Kant then attempts to unearth that which conditions the possibility of our experiencing ourselves as subjects inhabiting a world of objects. Alternatively, we might study the specific conceptual presuppositions of aesthetic judgements about entities as opposed to scientific hypotheses about them, or interrogate the distinctive presuppositions of the human sciences – the study of social and cultural structures and artefacts, and the guiding assumptions of those who investigate them as historians rather than as literary critics or sociologists.

In a terminology Heidegger sometimes employs in other texts, such ontological enquiries broadly focus on the what-being of entities[2] – their particular way or mode of being. Their concern is with what determines an entity as the specific type of entity it is, with that which distinguishes it from entities of a different type, and grounds both our everyday dealings with such entities and our more structured and explicit ontic investigations of the domain they occupy. Such a concern with what-being is to be contrasted with a concern with that-being. 'That-being' signifies the fact that some given thing is or exists,[3] and an ontological enquiry into that-being must concern itself with that which determines an entity of a specific type as an existent being – something equally fundamental both to our everyday dealings with it and to our ontic investigations of it, since neither would be possible if the entity concerned did not exist. A general contrast of this kind between what-being and that-being is thus internal to what Heidegger means by the Being of beings; it is a basic articulation of Being, something which no properly ontological enquiry can afford to overlook. And, indeed, the Western philosophical tradition since Plato has not overlooked it; but the way in which that tradition has tended to approach the matter has, for Heidegger, been multiply misleading.

With respect to the tradition's investigations of what-being, Heidegger will quarrel with the poverty and narrowness of its results. For, while human beings encounter a bewildering variety of kinds of entity or phenomena – stones and plants, animals and other people, rivers, sea and sky, the diverse realms of nature, history, science and religion – philosophers have tended to classify these things in ways that reduce the richness of their differentiation. The effect has been to impoverish our sense of the diversity of what-being, to reduce it to oversimple categories such as the Cartesian dichotomy between nature (*res extensa*) and mind (*res cogitans*) – a set of categories which, on Heidegger's view, obliterates both the specific nature of human beings and that of the objects they encounter. Similarly, the basic distinction between what-being and that-being has been subject to over-hasty and superficial conceptualizations. In medieval ontology, for example, it was taken up in terms of a distinction between essence (*essentia*) and existence (*existentia*) – a distinction which still has great influence over contemporary philosophical thinking, but which embodied a highly specific and highly controversial set of theological presuppositions, and which overlooks the possibility that the Being of certain kinds of entity (particularly that of human beings) might not be articulable in precisely those terms. And, of course, if this basic distinction has been improperly conceptualized, then the philosophical tradition's various attempts at comprehending the that-being of entities will have been just as erroneous as its attempts to grasp their what-being.

Accordingly, when Heidegger claims that the philosophical tradition has forgotten the question with which he is concerned, he does not mean that philosophers have entirely overlooked the question of the Being of beings. Rather, he means that, by taking certain answers to that question to be self-evident or unproblematically correct, they have taken it for granted that they know what the phrase 'the Being of beings' signifies – in other words, they have failed to see that the meaning of that phrase is itself questionable, that there is a question about the *meaning* of 'Being'. By closing off that question, they have failed to reflect properly upon a precondition of their ontological conclusions about the articulated unity of Being, and so failed to demonstrate that their basic orientation is above reproach; and this lack of complete self-transparency has led their investigations into a multitude of problems. As Heidegger puts it:

> The question of Being aims therefore at ascertaining the a priori conditions not only for the possibility of the sciences which examine entities as entities of such and such a type, and in so doing already operate with an understanding of Being, but also for the possibility of those ontologies themselves which are prior to the ontical sciences

and which provide their foundations. Basically, all ontology, no matter
how rich and firmly compacted a system of categories it has at its disposal,
remains blind and perverted from its ownmost aim, if it has not first
adequately clarified the meaning of Being, and conceived this clarification
as its fundamental task.

(BT, 2: 31)

RECLAIMING THE QUESTION

Nonetheless, apart from its earliest incarnation in ancient Greece, the philosophical tradition has tended to pass over this latter type of question in silence. As Heidegger begins his book by pointing out, 'this question has today been forgotten' (BT, 1: 21), largely because philosophers take themselves to have a multitude of reasons for dismissing it. Heidegger accordingly undertakes to counter each of those reasons; and, although he does so very briefly, the strategies he employs shed important light on his own, provisional understanding of what may be at stake in the question.

First, then, it might be argued that the question of the meaning of 'Being' can easily be answered; it is a concept just like any other, distinctive only in the sense that it is the most universal concept of all. In other words, Being is not a being, not a particular phenomenon we encounter in our active engagement with the world; rather, we arrive at our concept of it by progressive abstraction from our encounters with specific beings. For example, from our encounters with cats, dogs and horses, we abstract the idea of 'animalness'; from animals, plants and trees we abstract the idea of 'life', of 'living beings'; and then, from living beings, minerals and so on, we abstract the idea of that which every entity has in common – their extantness or being. What more need be said on the matter?

Heidegger is happy to accept the claim that Being is not a being; indeed, that assumption guides his whole project. He also accepts that our comprehension of Being is nonetheless bound up in some essential way with our comprehending interactions with beings. Being is not a being, but Being is not encounterable otherwise than by encounters with beings. For if Being is, as Heidegger puts it, 'that which determines entities as entities' (BT, 2: 25), the ground of their articulability in terms of what-being and that-being, then it is necessarily only to be met with in an encounter with some specific entity or other. In short, 'Being is always the Being of an entity' (BT, 3: 29). But he rejects the idea that Being relates to beings in the particular manner we outlined above; for the universality of 'Being' is not that of a class or genus, and so the term 'Being' cannot denote a specific realm of entities that might be placed at the very top of an ontological family tree. Membership of a class is standardly defined in terms of possession of a common property, but the 'members' of the 'class' of beings do not manifest such uniformity; the being of numbers, for example, seems not to be the same as the being of physical objects, which in turn differs from that of imaginary objects. In other words, if Being is not a being, neither is it a type or property of beings; it is neither a subject of predication nor a predicate.

Some philosophers have concluded from this that Being is undefinable: the very generality of the term 'Being', the fact that there is nothing – no entity or phenomenon – to which it does not refer, for them precisely demonstrates that there is nothing specific to which it does refer, that the term lacks any definable content. For Heidegger, however, this is a failure of philosophical imagination, an illegitimate leap from the perceived failure of a certain type of definition to the assumed failure of all types of explanation. The fact that 'Being' cannot be defined by delimiting the extension of a class shows only that a form of explanation suited to the analysis of entities and their properties is entirely unsuited to the clarification of 'Being'; it merely confirms that Being is neither an entity nor a type of entity. It does not show that some alternative clarificatory strategy, one that does not employ an inappropriate definitional template, could not shed some light on the matter.

Here, Heidegger cites approvingly Aristotle's suggestion that the unity of the realm of Being is at best one of analogy. He certainly does not think that this notion makes the meaning of Being completely transparent. But, by conceiving of the relation between mathematical entities, physical objects and fictional characters as a unity of analogy, Aristotle at least takes seriously our sense – evinced among other ways in an inclination to apply the term 'being' across such a variety of types of entity – of underlying interconnections between the various types of entity we meet, while avoiding the obviously mistaken preconceptions we rejected earlier. He thereby acknowledges the differences between the ontological structures grounding different domains of Being, without denying the possibility of uncovering a unified set of presuppositions grounding every such ontological structure. It is Aristotle's grasp of the articulated unity-in-diversity of Being – his sense of the categorial diversity implicit in our grasp of what-being, the categorial unity implicit in our grasp of that-being, and their mutual dependence – from which Heidegger wishes us to learn.

Anyone familiar with the work of Kant and Frege may, however, feel that Heidegger has so far succeeded only in making very heavy weather of relatively simple insights. For the Heideggerian claim that Being is neither an entity nor a property of entities might well bring to mind the lapidary phrase 'existence is not a real predicate' – often used to summarize the core of Kant's objection to the ontological proof of God's existence. If we claim that God is omnipotent, we predicate a property of a type of entity; we assert that entities of this – divine – type satisfy the conditions for application of the concept of 'omnipotence'. If, however, we claim that there is a God, we are not attributing the 'property' of existence to a type of entity but rather adding a type of entity to our tally of the furniture of the universe; in effect, we assert that the concept of a divine being does not lack application.

The difference is perspicuously captured in the Frege-inspired notation of first-order predicate calculus. Attributing existence to a type of entity is done by using the existential quantifier, rather than a predicate letter that corresponds to the putative 'property' of existence, in just the way that the letter 'O' might be used to capture the property of omnipotence, or the letter 'D' that of divinity. Thus, 'Any divine being is omnipotent' becomes: $\forall x [Dx \rightarrow Ox]$; whereas 'There is a [i.e. at least one] divine being' becomes: $\exists x [Dx]$. In other words, the supposedly mysterious and portentous meaning of Being, the significance of our use of the word 'is' to denote existence, is in fact fully captured in any competent explanation of the function of the existential quantifier.

We might think of this as a modern-dress version of the general claim that the meaning of Being is self-evident; and, once again, Heidegger would be happy to go along with some of its implications. It does, for example, provide one clear way of illustrating the claim that Being is not a property of beings, that the term is not a label for a specific class or type of entities. However, to think that invoking the elements of a logical notation is the best, or even the only, way of clarifying such a fundamental philosophical issue is to misunderstand the relation between logic and ordinary language.

The point of a logical notation such as the predicate calculus is to provide a perspicuous articulation of relations of deductive inference between propositions, thus permitting rigorous analysis of argumentative structures. This makes it a valuable tool for philosophical enquiry; but it means that the notation is designed to capture only one aspect of the propositions and arguments translated into it. Those aspects of the meaning of ordinary words and sentences deemed irrelevant to questions of deductive validity are simply lost in translation, leading to the usual warnings in logic textbooks that the propositional connectives associated with such terms as 'and' or 'if' must not be taken as synonyms for them. For example, if I claim that 'X hit Y with the baseball bat, and Y fell to the floor', I imply that the first event preceded and brought about the second; but an analysis of my claim that employs the conjunction sign '∧' carries no such implication. Given such discrepancies, however, why should we believe that the existential quantifier captures *every* aspect of the meaning of our term 'is' when it is employed to denote existence? On the contrary, we have good reason to believe that potentially crucial aspects of its meaning will not survive the translation into logical notation.

Moreover, even with respect to those aspects of linguistic meaning that logical notation does capture, why should we regard them as in any way philosophically trustworthy? In a logical notation, the propositions 'Peabody is in the auditorium' and 'Nobody is in the auditorium' will appear as symbolic strings with very different structures; but the precise form of those differences simply reflects our everyday understanding of the differences between the original propositions (e.g. the differences in the conclusions we can draw from their everyday utterance). In other words, our logical notation is only as good as our pre-existent, everyday understanding of our language, and so of the form of life in which it is ultimately grounded; and Heidegger will argue in *Being and Time* that that understanding is not to be trusted on matters of fundamental ontology. On the contrary: for Heidegger, as for many other philosophers, what seems obvious or most readily available to reflection may well lead us astray.

THE PRIORITY OF DASEIN

In short, Heidegger rejects the sorts of reasons standardly offered by philosophers for dismissing the question of the meaning of Being: it is neither unanswerable, nor possessed of a simple or self-evident answer. Nonetheless, that question has been systematically passed over in the discipline, to the point at which it now seems obscure and disorientating to most philosophers – and so to most of Heidegger's readers. Accordingly, before attempting to answer the question, an adequate or appropriate way of formulating it is required. We need to remind ourselves of what is involved in the asking of such a question – which means that we need to remind ourselves first of the fundamental structure of any enquiry, and then of this enquiry in particular.

Any enquiry is an enquiry about something. This means, first, that it has a direction or orientation of some sort, however provisional, from the outset; without some prior conception about what is sought, questioning could not so much as begin. Second, it means that any enquiry asks about something – the issue or phenomenon that motivates the questioning in the first place. In asking about this something, something else – some entity or body of evidence – is interrogated; and the result of its interrogation, the conclusion of the enquiry, is that something is discovered. But, most importantly of all, any enquiry is an activity, something engaged in by a particular type of being. It is thus something capable of being carried out in various possible ways – superficially or carefully, as an unimportant part of another task or as a self-conscious theoretical endeavour – all of which nonetheless must reflect, be understood as inflections of, the Being of the enquirer.

Seen against this template, certain distinctive aspects of our particular enquiry into the meaning of Being stand out. First, it is not a casual question but a theoretical investigation, one that reflects upon its own nature and purpose, attempting to lay bare the character of that which the question is about. But it, too, must be guided at the outset by some provisional, not-yet-analysed conception of what it seeks. Our questioning of the meaning of Being must begin (as ours did begin) within the horizon of a pre-existing but vague understanding of Being; for we cannot ask 'What *is* "Being"?' without making use of the very term at issue. There is, accordingly, no neutral perspective from which we might begin our questioning; the idea of a presuppositionless starting point, even for an exercise in fundamental ontology, must be rejected as an illusion. Our prior understanding of Being may well be sedimented with the distortions of earlier theorizing and ancient prejudices, which must of course be identified and neutralized as quickly as possible; but they can only be uncovered by unfolding that prior understanding with the utmost vigilance, not by avoiding contact with it altogether.

What of the threefold articulation of questioning that we laid out earlier? In our enquiry, that which is asked about (obviously enough) is Being – that which determines entities as entities, that on the basis of which entities are always already understood. Since Being is always the Being of an entity or entities, then what is interrogated in our enquiry will be entities themselves, with regard to their Being. And the hoped-for conclusion of the enquiry is – of course – the meaning of Being. But, if our interrogation is to deliver what we seek, then we must question those entities in the manner that is most appropriate to them and to the goal of our enquiry. We must find a mode of access to them that allows them to yield the characteristics of their Being without falsification.

We therefore need to choose the right entity or entities to interrogate, to work out how best to approach them, and to allow the real unity-in-diversity of Being to emerge thereby. In order to do these various things properly, we must clarify their nature and structure, make it clear to ourselves what counts as doing them well and doing them badly. But choosing what to interrogate, working out how to interrogate it, relying upon a preliminary understanding of

Being and attempting to clarify it: these are all modes of the Being of one particular kind of being, the kind for whom enquiring about entities with regard to their Being is one possibility of its Being – the entity which we are ourselves, the being Heidegger labels 'Dasein'. If, then, we are to pose our question properly, we must first clarify the Being of Dasein; it is from our everyday understanding of our own Being that we must attempt to unfold a more profound understanding of the question of the meaning of Being.

Heidegger's reasons for introducing the term 'Dasein' – which, translated literally, simply means 'there-being' – where it would seem natural to talk instead about human beings, are manifold. First, in everyday German usage, this term does tend to refer to human beings, but primarily with respect to the type of Being that is distinctive of them; it therefore gives his investigation the right ontological ring. Second, it permits him to avoid using other terms that philosophers have tended to regard as synonymous with 'human being', and have concentrated upon to the point at which they trail clouds of complex and potentially misleading theorizing. Time-hallowed terms such as 'subjectivity', 'consciousness', 'spirit' or 'soul' could only be prejudicial to Heidegger's enquiry. Third, and consequently, an unusual term such as 'Dasein' is a tabula rasa: devoid of misleading implications, it can accrue all and only the significations that Heidegger intends to attach to it. The rest of Heidegger's analysis of the Being of Dasein is thus, in effect, an extended definition of its core meaning – a working-out of the furthest implications of the intentionally uncontroversial assumption that human beings are beings who ask questions.

With these words of warning, we can return to Heidegger's main line of argument. He has already identified Dasein as the object of an enquiry that must *precede* any proper posing of the question of the meaning of Being. But he also claims that Dasein is the most appropriate entity to be interrogated in the posing of that question, i.e. that working out an ontological characterization of Dasein is not just an essential preliminary to, but forms the central core of, fundamental ontology. In so doing, Heidegger makes certain claims about the Being of human beings, claims that can only be fully justified and elaborated in the body of *Being and Time*, but which must at least be sketched in here. First, and most importantly, then, Dasein is said to be distinctive among entities in that it does not just occur; rather, its Being is an issue for it. What might this mean?

All entities exist in the sense that they are encounterable in the world; some exist in the sense that they are alive; but, of them, only Dasein exists in the sense that the continued living of its life, as well as the form that its life will take, is something with which it must concern itself. Glasses and tables are not alive at all. Cats and dogs are alive but they do not have a life to lead: their behaviour and the ways in which they encounter other entities (as harmful, satiating, productive of pleasure and pain) are determined by the imperatives of self-preservation and reproduction; they have no conscious, individual choice as to how they want to live, or whether they want to continue living at all. Only human creatures *lead* their lives: every impending moment or phase of their lives is such that they have it to be, i.e. they must either carry on living in one way or another, or end their lives. Although this practical relation to one's existence can be repressed or passed over, it cannot be transcended; for refusing to consider the questions it raises is just another way of responding to them, a decision to go on living a certain kind of life. After all, if Dasein is the being who inquires into the Being of all beings, the same must be true of its relation to its own Being; its existence necessarily confronts it with the question of whether and how to live. In Heidegger's terms, Dasein's own Being (as well as that of other beings) is necessarily an issue for it.

The Being of Dasein cannot, then, be understood in the terms usually applied to other types of entity; in particular, we cannot think of Dasein as having what we have called what-being, a specific essence or nature that it always necessarily manifests. Such terms are appropriate to physical objects and animals precisely because how and what to be is never a question for them; they simply are what they are. But, for Dasein, living just *is* ceaselessly taking a stand on who one is and on what is essential about one's being, and being defined by that stand. In choosing whether or not to work late at the office, to spend time with the family, to steal a purse, to travel to a rock concert, one chooses what sort of person one is.

In identifying with certain activities, character traits, life styles and visions of the good and in rejecting others, we reveal our conception of what it is to flourish as a human being, and so of what it is to *be* a human being and make it concrete in our own existence.

In so doing, of course, the precise nature and array of physical and mental capacities that human beings possess, and their natural impulses towards self-preservation and reproduction, must be taken into account; but just *how* a given individual does so – how she interprets their significance – remains an open question. The human way of Being is not simply fixed by species-identity, by membership of a particular biological category; Dasein is not *homo sapiens*. Similarly, the array of lifestyles and interpretations of human possibilities and human nature available in our culture will set limits on our capacity for self-interpretation (becoming a Samurai warrior is simply not a possibility for a citizen of early twenty-first-century London). But which feasible self-interpretation is chosen and how it is adapted to person-specific circumstances remains an issue for each individual; and, since each choice, once made, could be unmade or otherwise adapted in the future, each new moment confronts us with the question of whether or not to stick with choices already made. Hence, the issue of one's existence is never closed until one no longer exists.

One could conclude that Dasein's essence must lie in this capacity for self-definition or self-interpretation, and in one sense this would be right, since that is what most fundamentally distinguishes Dasein from other entities. It would be misleading, however, for this particular capacity is unlike any of those used to define the what-being of other entities; its exercise fixes who and what the entity is, rather than being one manifestation of the entity's already fixed nature. It seems better to stick with Heidegger's formulations, namely that Dasein's essence lies in existence, that for it alone existence is a question that can be addressed only through existing, and so that it alone among all entities can be said properly to exist. In line with this, he invites us to think of the particular self-interpretation that a given Dasein lives out, the existential possibility it chooses to enact, as an *existentiell* understanding, which he regards as determining its ontic state; and he thinks of any ontological analytic of

Dasein, any attempt to uncover the structures which make any and all existentiell understandings possible, as an *existential* analytic.

This distinctive characteristic of Dasein will be examined in more detail later,[4] but we can already see why Heidegger thinks that Dasein is the type of entity which must be interrogated in any exercise in fundamental ontology. For the aim of any such exercise is to interrogate Being as it makes itself manifest through the Being of an entity; and the fact that Dasein's essence is existence makes the relationship of its Being to Being a peculiarly intimate one in at least three respects. First, unlike any other entity, every ontic or existentiell state of Dasein embodies a relationship to its own Being – in so far as it exists, every Dasein relates itself to its own Being as a question. Second, every such relationship embodies a comprehending grasp of its Being – a particular answer to the question that its Being poses; its every existentiell state is thus implicitly 'ontological', making manifest an undertanding of Dasein in its Being, and so an understanding of Being. Third, in enacting any given existentiell state Dasein necessarily relates itself to the world of entities around it. I can't take a shower or read a thriller without engaging with the tools of my chosen project; so Dasein is always already relating itself comprehendingly (and questioningly) to other entities as the entities they are, and as existent rather than non-existent. 'Dasein has therefore a third priority as providing the ontico-ontological condition for the possibility of any ontologies' (BT, 4: 34).

Given this threefold priority of Dasein, the provision of an existential analytic of Dasein would inevitably provide the richest, most complete and so most revelatory way of engaging in fundamental ontology. Being is only encounterable as the Being of some entity or other, and entities come in a bewildering variety of forms. So, if the fundamental ontologist chooses to interrogate an entity other than Dasein, she will emerge at best with a deeper grasp of the Being of that kind of being alone; and then the task of grasping Being as such or as a whole will seem – impossibly – to require that she interrogate every specific kind of being in its Being, in order to combine the individual results. But if she can understand the Being of Dasein, the only entity for whom Being as such is an issue, she will grasp what it is for an entity to relate itself comprehendingly and questioningly towards the Being of any and every entity (including itself), i.e. towards any manifestation of Being whatever. She will, in other words, acquire an understanding of what it is to understand Being; and since what is understood in an understanding of Being is indeed Being, to grasp the constitutive structure of that understanding (that which permits it to take the Being of any and all beings as its object) will be to grasp the constitutive structure of that which is thereby understood (what it is for Being, in any and every one of its ways, shapes and forms, to 'be'). As I suggested earlier, then, an existential analytic of Dasein is not merely an essential preliminary to the task of fundamental ontology; rather, 'the ontological analytic of Dasein in general is what makes up fundamental ontology' (BT, 4: 35).

PHILOSOPHY, HISTORY AND PHENOMENOLOGY

Having determined the appropriate object of interrogation for his enquiry, Heidegger then outlines the way in which he proposes to approach it. He does not, for example, want his enquiry to be guided by the most obvious or widely accepted, everyday understanding of Dasein's Being. Since Dasein's own Being is an issue for it, it always operates within some particular understanding of its own Being, and in that sense Heidegger's enquiry is simply the radicalization of a tendency that is essential to Dasein's Being. But it doesn't follow that the self-understanding with which Dasein's ordinary modes of existence are imbued will provide a fundamental ontological investigation with its most suitable orientation; for all we know at this stage, radicalizing that self-understanding may ultimately involve reconstructing it from the ground up. Neither, however, does Heidegger want to rely upon the deliverances of any ontic science: although Dasein's nature and behaviour have been studied over the years by a multitude of disciplines, we have no guarantee that the existential underpinnings of their existentiell investigations were reliably derived from Dasein's true nature, rather than from dogmatically held theoretical prejudices rendered 'self-evident' solely by the cultural authority of a particular ideological tradition or philosophical school.

We need, therefore, to return to the object of interrogation itself, unmediated (as far as that is possible) by already existing accounts and theories; and we need to study it in resolutely non-specialized contexts, in order to avoid assuming that aspects of this entity's behaviour or state that are specific to such atypical situations are in fact manifestations of its essential nature. For Heidegger, this means that Dasein must be shown 'as it is *proximally and for the most part* – in its *average everydayness*. In this everydayness there are certain structures which we shall exhibit – not just any accidental structures but essential ones which, in every kind of Being that Factical Dasein may possess, persist as determinative for the character of its Being' (BT, 5: 37–8). Heidegger is not assuming that Dasein's ordinary or usual state is the one that most fully and authentically expresses Dasein's possibilities – any more than he is inclined to rely upon the self-understanding that informs that state; as we shall see, he thinks that precisely the reverse is the case. But he does think that this state, like any other state of Dasein, must manifest those structures that are constitutive of its Being; and the philosophical tradition's tendency to overlook or ignore it makes it more likely that we will be able to characterize it in a way that is not distorted by misleading preconceptions. The realm of the ordinary is thus our best starting point; it may not provide the last word on the philosophical issues with which we are concerned, but it can and ought to provide the first.

Nevertheless, no enquiry into Dasein's average everydayness can begin without a preliminary conception of its overall goal or purpose, and of the specific aspects of the object of interrogation that will prove to be most illuminating or revelatory. As we saw earlier, a truly presuppositionless enquiry would lack all direction. If, however, this enquiry is to be completely transparent to itself and to those reading its results, its preconceptions must be explicitly declared and acknowledged. Accordingly, Heidegger announces that 'we shall point to *temporality* as the meaning of the Being of that entity which we call "Dasein"' (BT, 5: 38). His existential analytic will attempt to show that the constitutive structures of Dasein must ultimately be interpreted as modes of temporality, and that, consequently, whenever Dasein tacitly understands something like Being (whether its own or that of any other entity), it does so with time as its standpoint. If, however, all ontological understanding is rooted in time, it follows that the meaning of Being cannot be understood except in terms of temporality, against the horizon of time. 'In the exposition of the problematic of Temporality the question of the meaning of Being will first be concretely answered' (BT, 5: 40).

We must, of course, wait until this programme is carried out in detail before attempting to evaluate its success and its significance; but this preliminary declaration is indispensable for understanding the approach that Heidegger will adopt in the first stage of his enquiry – his provision of an existential analytic of Dasein. For engaging in such an enquiry is itself an ontical possibility of Dasein, an endeavour that only Dasein among all entities is capable of carrying out; so its basic structure must necessarily conform to the limits set by Dasein's existential constitution. And if that constitution is essentially temporal, then any enquiry into that constitution ought to understand itself as rooted in time, and so as historical in a very specific sense. Rocks and plants have a history in the sense that they have occupied space and time for a certain period during which certain things have happened to them. Dasein, however, exists; it leads a life in which its own Being is an issue for it. But, then, events in its past cannot be thought of as having been left behind it, or at most carried forward as memories or scars. Dasein does not merely have a past but lives its past, it exists in the terms that its past makes available for it – the question that its Being poses for it is always and ineliminably marked by its historical circumstances. As Heidegger puts it:

> Whatever the way of being it may have at the time, and thus with whatever understanding of Being it may possess, Dasein has grown up *into* and *in* a traditional way of interpreting itself: in terms of this it understands itself proximally and, within a certain range, constantly. By this understanding, the possibilities of its Being are disclosed and regulated.
>
> (BT, 6: 41)[5]

If, however, this is generally true of Dasein, it must also be true of Dasein as an ontological enquirer. Heidegger's preliminary understanding of Dasein therefore commits him to understanding his own enquiry as emerging into a tradition of ontological enquiry, and so as attempting to advance that tradition, to project it into the future; but also as ineliminably marked by the history of that tradition, as the place in which that history is lived out in the present. This inherent historicality has many implications. First, it means that Heidegger is attempting to pose a question whose true significance has been doubly distorted over the centuries. On the one hand, the tradition of ontological enquiry has attempted to cover up or pass over the question of the meaning of Being altogether; and, on the other, it has developed ontological categories in terms of which to understand specific regions of Being that have come to appear as self-evident and so as effectively timeless deliverances of reason (here, Heidegger has in mind such notions as Descartes' *ego cogito* or the Christian conception of the soul as categories for understanding Dasein). If, therefore, Heidegger's question is to be answered properly, he must break up the rigid carapace with which this tradition confronts him. He must find a way of posing it that recovers its profundity and difficulty; and he must reveal the historical contingency of seemingly self-evident, philosophical categorizations of various types of entity, show that these 'timeless' truths are in fact the fossilized product of specific theorists responding to specific, historically inherited problems with the specific resources of their culture.

Heidegger does not, however, regard the philosophical tradition purely as something constraining or distorting. What he inherits from the past, that which defines and delimits the possibilities with which he is faced in engaging with his fundamental question, is not simply to be rejected. After all, the complete and undiscriminating rejection of every possibility that his tradition offers would leave him with no orientation for his enquiry, with no possible way of carrying on his questioning. In fact, the philosophical past with which he must live is a positive inheritance in two central respects. First, if Dasein's understanding of Being is constitutive of its Being, then it can never entirely lose that understanding. It must, therefore, be possible to recover something potentially valuable for an ontological enquiry from even the most misleading and distortive theoretical systems of the philosophical tradition. And, second, Heidegger never claims that every contribution to this tradition was benighted; on the contrary, he stresses the positive elements of relatively recent philosophical work (such as Kant's emphasis upon time as a form of sensible intuition), and he places particular emphasis upon the value of work done at the very outset of this tradition in ancient Greece (unsurprisingly, since if such work did not contain a fundamentally sound initial grasp of the question of the meaning of Being, nothing resembling a tradition of ontological enquiry could have originated from it).

Thus, Heidegger's persistent concern with the historical matrix of his existential analytic is not just a scholarly and dramatic but essentially dispensable way of illuminating issues that might easily be examined in other ways; it is the only way in which this kind of enterprise can find its proper orientation and grasp the most fruitful possibilities that are available to it. There can be no fundamental ontology without the history of fundamental ontology, no philosophy without the history of philosophy. And Heidegger's conception of the relationship of his own enquiry to its history is neither simply negative nor simply positive: it is neither destruction nor reconstruction, but rather deconstruction. It thus forms the point of origin of the recently popular and controversial strategies in the human sciences that have come to be known by that label, and that are perhaps most often associated with the name of Derrida. It may be that, if we relate Derrida's work to its (often explicitly acknowledged) Heideggerian origins, we might come to see that its relation to the history of philosophy is no less nuanced and complex than Heidegger's own; in other words, we might appreciate that deconstruction is not destruction.

But, if deconstruction is one inheritor of Heideggerian fundamental ontology, and is one of the future possibilities it opens up for the discipline of philosophy, its most immediate ancestor – that element of the philosophical past of which Heidegger deems his work to be the living present – is Husserlian phenomenology. Given Heidegger's own sense of the need to understand the immediate circumstances of a theory's production if one is to grasp its most profound insights and errors, it would seem essential to comprehend the Husserlian background of his own enquiry. However, when, at the very end of his introduction to *Being and Time*, he claims the title of 'phenomenology' for his work, he acknowledges Husserl's influence and originality but deliberately fails to provide any detailed analysis of his relation to the Husserlian project. Instead, he offers an etymological analysis of the term itself, and derives his own project therefrom.

This omission (or, better, displacement) is a puzzle.[6] But it would be foolhardy to assume in advance that the mode of derivation with which readers of *Being and Time* are confronted is inadequate for its author's purposes. On the contrary, the most appropriate interpretative principle to adopt must surely be that Heidegger's decision in this respect has an internal rationale – that it gives him precisely what he perceives to be required, and does so in a more satisfactory manner than any alternative available to him. Only if this assumption turns out to generate a manifestly inadequate interpretation of the book as a whole can it be justifiable to turn our attention to issues that its author excluded from the text itself. Accordingly, I intend to observe Heidegger's own circumspection, and concentrate on the central points that his employment of the label 'phenomenology' in *Being and Time* itself seems intended to highlight.

First, Heidegger asserts that 'phenomenology' names a method and not a subject matter. It is therefore unlike its cousins 'theology' or 'methodology', which offer an articulated, systematic account of what is known about a particular type of entity, region or mode of Being. Phenomenology, according to Heidegger, does not demarcate any such region: it

> expresses a maxim which can be formulated as 'To the things themselves!' It is opposed to all free-floating constructions and accidental findings; it is opposed to taking over any conceptions which only seem to have been demonstrated; it is opposed to those pseudo-questions which parade themselves as 'problems', often for generations at a time.
>
> (BT, 7: 50)

Unfortunately, this seems little more than a set of empty platitudes. No one is likely to declare themselves in favour of pseudo-questions or free-floating constructions; the issue is how one might best avoid them. However, Heidegger provides a more precise definition of his method by etymological means – by analysing the two semantic elements from which the term 'phenomenology' has been constructed, namely 'phenomenon' and 'logos'. What matters most for our purposes, of course, is not the accuracy of these derivations, but what is derived from them.

We will take 'logos' first. As Heidegger points out, this Greek term is variously translated as 'reason', 'judgement', 'concept', 'definition', 'ground' or 'relationship' (and we might add to this 'law' and 'word' – or 'Word', as the term is translated in the Prologue to St John's Gospel). He claims, however, that its root meaning is 'discourse' – but 'discourse' understood not as 'assertion' or 'communication', but as 'making manifest what one is "talking about" in one's discourse' (BT, 7: 56). For the fundamental aim of discursive communication is to communicate something about the topic of the discourse; what is said is, ideally, to be drawn from what is being talked about, and to be displayed as it truly is. More modern emphases upon truth as a matter of agreement or correspondence between judgement or assertion and its object fail to consider what must be the case for such agreement to be possible. In particular, they fail to see that a judgement can only agree or disagree with an object if the object has already been uncovered or discovered in its Being by the person judging. This is no more than a sketch of an argument that Heidegger will develop later in his book; so its validity can hardly be assessed here.[7] Nevertheless, it is this fundamental uncovering or unconcealing of entities in their Being to which he claims that the Greek term 'logos' originally refers; and it is this with which the phenomenologist concerns herself.

A similar significance is held to accrue to the Greek term 'phenomenon', on Heidegger's account of the matter. Here, the point that we must bear in mind is that 'the expression "*phenomenon*" signifies *that which shows itself in itself*, the manifest. Accordingly . . . phenomena are the totality of what lies in the light of day or can be brought to the light' (BT, 7: 51). Of course, entities can show themselves in many different ways: they may appear as something they are not (semblance), or as an indication of the presence of something else that does not show itself directly (symptoms), or as the manifestation of something that is essentially incapable of ever manifesting itself directly (the Kantian idea of phenomena as opposed to noumena, of the content of empirical intuition understood as an emanation of the necessarily non-encounterable thing-in-itself). The distinctions between these different kinds of appearances are important; but they all show themselves in themselves, in accord with their true nature, and so they all count as phenomena in the formal, root sense Heidegger identifies.

However, the phenomenological sense of the term 'phenomenon' is more specific than this. It is best illustrated by an analogy with an element of Kant's theory of knowledge, within which space and time are conceived as forms of sensible intuition. According to Kant, space and time are neither entities nor properties of entities, and so not discoverable as part of the content of sensible intuition; but our experience of the world is only possible on the assumption that the objects we thereby encounter occupy space and/or time, i.e. on the assumption that experience takes a spatio-temporal form. On this account, space and time constitute the horizon within which any object must be encountered, and so in a certain sense necessarily accompany every such entity; but they are not themselves encounterable as objects of experience and neither are they separable components of it. A sufficiently self-aware and nuanced philosophical investigation of their status, however, can make them the object of theoretical understanding, and thus thematize what is present and foundational but always unthematized in everyday experience.

Heidegger defines the 'phenomena' of phenomenology in terms that suggest that they occupy a place in human interactions with entities that is strongly analogous to the Kantian conceptions of space and time:

> That which already shows itself in the appearance as prior to the 'phenomenon' as ordinarily understood and as accompanying it in every case, can, even though it thus shows itself unthematically, be brought thematically to show itself; and what shows itself in itself (the 'forms of the intuition') will be the 'phenomena' of phenomenology.
>
> (BT, 7: 54–5)

The Kantian analogy makes it clear that the 'phenomena' of phenomenology are not appearances in any of the three senses we distinguished above; for the forms of sensible intuition do not appear as what they are not, and they are not signs of something else that is or must be non-manifest. But neither are they something necessarily non-manifest; for space and time can be brought to show themselves as what they are by the Kant-inspired philosopher, and accordingly not only count as phenomena in the formal sense of that term, but also as a fit subject for discourse or 'logos' in the root sense of that term, and so for phenomenology itself.

But these considerations tell us only what the object of phenomenology is not; they shed no light on what it is. What exactly is a 'phenomenon' in the phenomenological sense?

> Manifestly, it is something that proximally and for the most part does *not* show itself at all: it is something that lies *hidden*, in contrast to that which proximally and for the most part does show itself; but at the same time it is something that belongs to what thus shows itself, and it belongs to it so essentially as to constitute its meaning and its ground. Yet that which remains *hidden* in an egregious sense, or which relapses and gets *covered up* again, or which shows itself only '*in disguise*', is not just this entity or that, but rather the Being of entities, as our previous observations have shown. This Being can be covered up so extensively that it becomes forgotten and no question arises about it or about its meaning.
>
> (BT, 7: 59)

If 'phenomenology' has to do with the logos of phenomena, if it lets that which shows itself be seen from itself in the very way in which it shows itself from itself, then it is and must be our way of access to the Being of entities – its meaning, modifications and derivatives. Fundamental ontology is possible only as phenomenology: only that method fits that subject matter. Phenomenology is the science of the Being of entities.

CONCLUSION: HEIDEGGER'S DESIGN

We can now see how Heidegger's preliminary reflections on the proper form of his enquiry into the meaning of Being delivered the specific plan for its treatment that we find at the end of the Introduction to *Being and Time*. Since Being is always the Being of an entity, any such enquiry must choose one particular type of entity to interrogate, and locate the most appropriate means of access to it. Since such an enquiry is a mode of Dasein's Being, it can be fully self-transparent only if preceded by an existential analytic of Dasein. But Dasein's Being is such that its own Being is an issue for it and it can grasp the Being of entities other than itself. Such a peculiarly intimate relationship with Being in all its manifestations implies that an existential analytic of Dasein should also form the centrepiece of that enquiry. That existential analytic will reveal that the constitutive structures of Dasein's Being are modes of temporality; and since Dasein is the ontico-ontological precondition for any understanding of Being, time must be the horizon for understanding the meaning of Being. But if Dasein's Being is essentially temporal, the enquiry which reveals this must itself be essentially historical, a living-out in the present of the tradition of philosophical investigations into Being. It must therefore free itself for a fruitful future by deconstructing its own history – rescuing the question of Being from oblivion, revealing the historically specific origins of seemingly timeless interpretations of Being and beings, and recovering their more positive possibilities.

Accordingly, Heidegger's project falls into two parts, each consisting of three divisions. In the first part, an existential analytic of Dasein is provided (Division One), which is then shown to be grounded in temporality (Division Two), and time is explicated as the transcendental horizon for the question of Being (Division Three). In the second part, a phenomenological deconstruction of the history of ontology is worked out by means of an investigation of Kant's doctrine of schematism and time (Division One), Descartes' *ego cogito* (Division Two), and Aristotle's conception of time (Division Three). In reality, however, only the first two divisions of Part One were originally published under the title *Being and Time*, and the missing divisions were never added in subsequent reprintings. In other words, Heidegger's *magnum opus* contains only his interpretation of Dasein's Being in terms of temporality.

This fact about the book – its status as part of a larger whole – is absolutely critical to a proper understanding of it, but it requires very careful handling. Placing undue stress upon the scope of Heidegger's original design for the book can contribute to a profound misreading of it; for our attention can thereby be focused upon the mismatch between intention and execution in such a way as to imply that, because *Being and Time* is an unfinished book, the larger project adumbrated in its opening pages was also left uncompleted. Speculation then abounds concerning the reasons for this lack of closure. Does it mean that Heidegger simply never got around to working out what he wished to say under the four missing general headings, or rather that he came to realize that those elements of his project, and so the wider project as a whole, were fundamentally unrealizable?

However, it is simply wrong to assume – as such speculation presupposes – that the other four divisions of *Being and Time*, or at least a set of texts whose manifest topic and general methodological spirit approximate to them, are unavailable. Heidegger published his detailed analysis of Kant's doctrine of schematism and time as a separate book (*Kant and the Problem of Metaphysics*)[8] in 1929. His explication of time as the horizon of the question of Being, together with an investigation of Cartesian ontology and the Aristotelian conception of time, were made public in the form of lectures at the University of Marburg in 1927 (the year of *Being and Time's* publication), and have now been published under the title *The Basic Problems of Phenomenology*.[9] If we put these three volumes together, then, we have the entire treatise that Heidegger had originally wished to call '*Being and Time*' – even if not in the precise form he envisaged.[10] Although, therefore, Heidegger may later have come to believe that his initial conception of the task of philosophy was in some ways inadequate, it is wrong to think that he abandoned its execution at the point at which the extant text of *Being and Time* ends.

The existence of these complementary texts also deprives us of any excuse for failing to read *Being and Time* as part of a wider project; it acts as a salutary reminder that, if we must not over-interpret the fact that *Being and Time* is unfinished, neither must we underplay it. In particular, we must not take the de facto separation between Divisions One and Two of Part One and Division

Three as evidence of a conceptual or methodological separation between the work done in these two places; for Heidegger always understood his existential analytic of Dasein to be part of his wider enquiry into the meaning of Being. The exclusive focus of *Being and Time* upon the Being of Dasein is thus not a sign that Heidegger's understanding of his central project is anthropocentric – at least in any obvious or simple way. His primary concern is always with the question of the meaning of Being; so we must never forget that what we know as *Being and Time* comes to us in a significantly decontextualized form.

One final word of warning is in order concerning the sense in which *Being and Time* is an unfinished work. It is at least possible that the unfinished appearance of the text is in fact deceptive, a function of the expectations with which we approach it rather than a reflection of its true condition. By presenting us with a text that appears to be incomplete, it may be that Heidegger is attempting to question our everyday understanding of what is involved in completing a philosophical investigation – of what it might mean to bring a line of thought to an end. After all, he certainly questions our everyday understanding of how a philosophical investigation should *begin*; on his account, no type of human enquiry can conceivably take the essentially presuppositionless form that is often held up as the ideal for philosophical theorizing. And if Dasein's comprehending grasp of beings in their Being is always a questioning one – embodying an understanding that is not only the result of prior questioning, but that will itself engender further questions, and hence always be open to modification – then Dasein could not conceivably attain an understanding of anything that was beyond any further question. So the very idea of an absolutely final result of human inquiry makes no more sense for Heidegger than that of an absolutely pure starting point; both the origins and the termini of a temporal being's questioning cannot be other than conditioned and conditional.

It would therefore be the very reverse of surprising to discover that the concluding pages of *Being and Time* – with their air of incompletion, their references to work as yet undone, and their emphasis upon reformulating questions rather than providing definitive answers to them – are as conclusive, as exemplary, of what it is to achieve a terminus in philosophy, as could coherently be desired. For the idea that a philosophical project is complete only when it has definitively answered all the questions it sets itself, and the idea that a text is complete only when it no longer calls for its own continuation, are not so much ideals to which all philosophers should aspire as illusions with which they must learn to dispense. We will return to this issue in the concluding pages of this book; but readers should bear in mind from the outset that *Being and Time*'s seemingly self-evident failure to carry through the task it sets itself does not necessarily mean that its philosophical work is incomplete.

Before we turn to an examination of that work, however, I want to stress what is philosophically distinctive about Heidegger's conception of his general project. His focus upon the particular nature of human existence is not, of course, unusual in the history of philosophy; particularly in the modern period, it has been absolutely central to the discipline. What is unusual, however, is the wider framework of Heidegger's analysis. Indeed, the very idea that there might be such a thing as a question about Being itself, one which underlies any questions about specific regions of Being and their ontological underpinnings, is one that Heidegger needs to rescue from oblivion before he can work towards any sort of answer to it. And this involves him in the salutary task of getting his readers to see that a question can be asked at a level that is normally immune from interrogation. Philosophers typically force non-philosophers to ask questions that disrupt the assumptions upon which their everyday activities are based; sceptical problems about induction and other minds exemplify this to perfection. It is therefore intriguing, and potentially educative, to see the same procedure directed at the unthinking assumptions of philosophers themselves. Even if, in the end, we were to dismiss Heidegger's question, his attempts to raise it would at least have forced us to reflect upon something we otherwise take for granted.

It is this sort of heightened self-awareness that is the most distinctive aspect of Heidegger's work; his investigation is permeated with an awareness of its own presuppositions. First, he makes explicit from the outset the preconceptions about his subject matter that are orienting his analysis; they are not left in obscurity to be unearthed by disciples and exegetes, but are themselves made the subject of analysis – an analysis which identifies the essential role of such preconceptions in any enquiry. Second, he is sensitive to the fact that his enquiry forms one part of a long tradition of philosophical endeavour, from which in part it inevitably derives its orientation and which necessarily furnishes him with tools and traps – with essential conceptual resources and rigidified, seemingly self-evident categorizations. Perhaps more than any other philosopher (Hegel excepted), Heidegger understands that the present, and so the future, of his subject cannot be understood apart from its history, that the history of philosophy belongs to philosophy and not history; he works in the knowledge that all such work can be fruitful only by acknowledging its past. Third, Heidegger writes in the constant awareness that such writing is a human act, the enactment of a human possibility; he is a being whose ways of being are the subject of his work, so its results must feed back into and inform its conduct.

The implications of this last point are multiple and profound. To begin with, it suggests an important methodological principle for this and any other discipline whose topic is Dasein. Only an enquiry that is informed by the richest and most accurate understanding of what it is for Dasein to exist as an enquirer can itself be rich and accurate; but that understanding can only be achieved by an enquiry into Dasein's Being. For Heidegger, this does not spell contradiction – with the enquirer into Dasein unable to begin until she finishes; it reveals the existence of what is called the hermeneutic circle in the human sciences. Its implication is not that beginning an enquiry is impossible, but that it cannot be presuppositionless; accordingly, presuppositions ought not to be eschewed, but rather acknowledged and used to best effect. We must enter the circle by initiating our enquiry on the basis of some preconception (provisional, but worked out with maximal care) and, then, when we reach a provisional conclusion, return to our starting point with the benefit of a deeper understanding, which can then render one's next set of conclusions more profound – and so on around the circle. This is one reason why Division Two of *Being and Time* works over again the material generated by Division One, deepening its insights on Heidegger's second tour of his own particular circuit.

This awareness of the humanity of all enquirers into Dasein and the meaning of Being leads to a second important methodological principle – the need for a diagnostic element in philosophical criticism. For Heidegger claims *both* that Dasein is the being uniquely possessed of an understanding of Being, and that its enquiries into Being constantly and systematically misunderstand it – claims which together imply that Dasein is constantly and systematically out of tune with that with which it is nonetheless most fundamentally attuned. Such a persisting and fundamental misalignment, an incomprehension that is not merely intellectual but must rather inform Dasein's existentiell states, clearly requires explanation. And seen against this background, Heidegger's own avowed ability to avoid those errors, to perceive the grains of truth in seemingly self-evident traditional categorizations and to resurrect and reorient enquiries into Being, itself needs accounting for. How can he see what so many others have missed and persist in missing? In other words, in Heidegger's philosophy, philosophical misunderstandings call not only for identification but for the provision of an aetiology, a diagnosis of how and why the human beings who elaborated them might have gone wrong about something so close to their own natures.

And the necessary diagnostic tools are provided by the existential analytic of Dasein itself. For Heidegger, because Dasein's Being is such that its own Being is an issue for it, any given mode of its existence can be assessed in terms of what he calls authenticity or inauthenticity. We can always ask of any given individual whether the choices she makes between different possible modes of existence and the way she enacts or lives them out are ones through which she is most truly herself, or rather ones in which she neglects or otherwise fails to be herself. The full significance of this terminology will emerge in the following chapter; but if its general pertinence to human life can be properly established, it must apply to the way in which individuals have prosecuted the specific task of enquiring into the meaning of Being. If philosophers have not done so in the most authentic possible way, if they have not properly seized upon such enquiry as an existentiell state of their Being, their results will be correspondingly inauthentic. As Heidegger puts it:

> the roots of the existential analytic . . . are ultimately *existentiell*, that is, *ontical*. Only if the enquiry of philosophical research is itself seized upon in an existentiell manner as a possibility of the Being of each existing Dasein, does it become at all possible to disclose the existentiality of existence and to undertake an adequately founded ontological problematic.

> (BT, 4: 34)

This is Heidegger's basic diagnostic assumption about the errors of his predecessors and his colleagues: their failure to pose the question of Being correctly is caused by and is itself a failure of authenticity. It follows, of course, that the task of posing it correctly will only be achievable by an existentially authentic enquirer. Heidegger has the arrogance to think that this is what he has at least begun to achieve: but he has the humility to know that any errors he accrues along the way will reveal his own inauthenticity. And his achievement, if it is indeed real, is one which will not benefit him alone; for what he then offers to his readers in his existential analytic is at once the means to diagnose their own inauthenticity and the means to overcome it. Indeed, in the course of this book it will gradually become clear that the work Heidegger intends to accomplish in *Being and Time* can only be understood if we appreciate his constant attentiveness to the relationship that his words at once allow him and compel him to establish and maintain with his readers.

To invoke questions of authenticity within the precincts of philosophical endeavour was once a commonplace: to engage in philosophizing was long understood as a way, perhaps *the* way, of acquiring wisdom about the meaning of human existence, and thus of leading a better life. Nowadays, the idea that one's success or failure at philosophizing can legitimately be assessed at all in personal terms is not often considered; and the idea that one's philosophical position might be criticized as existentially inauthentic might appear either ludicrous or offensive. Such reactions betoken a conception of the subject that represses the fact that it is human beings who produce philosophy, that philosophizing is a part of a human way of living. It is, of course, perfectly possible to act out such a repression; nothing is easier than to write philosophy in a way that represses the fact of one's own humanity. But, as Kierkegaard pointed out, such forgetfulness – particularly when one's very topic is what it is to be human – is liable, where it is not comic, to be tragic in its consequences. In *Being and Time*, Heidegger attempts to trace out the tragi-comic effects of this repression in the history of the subject, and to demonstrate the fertility and power that is released when that repression is lifted.

NOTES

1 W. Golding, *The Spire* (London: Faber and Faber, 1964).
2 See the Introduction to *The Basic Problems of Phenomenology*, trans. A. Hofstadter (Bloomington, Ind.: Indiana University Press, 1982), p. 18. At BT, 2: 26. Heidegger uses the term 'Sosein' (translated as something's 'Being as it is') to gesture towards a broadly similar idea.
3 See the reference to *The Basic Problems of Phenomenology* in note 2. At BT, 2: 26, Heidegger uses the term 'Daß-sein' (translated as 'the fact that something is') to pick out this aspect of the Being of beings.
4 See particularly Chapter 2. Some readers will already have detected that this account of Heidegger's conception of Dasein bears a close family resemblance to Charles Taylor's explicitly Heideggerian account of human beings as self-interpreting animals. Taylor works out the details of this account in various places: see particularly his 'Interpretation and the Sciences of Man' and 'Self-interpreting Animals' (in *Philosophical Papers* [Cambridge: Cambridge University Press, 1985]), and Part One of *Sources of the Self* (Cambridge: Cambridge University Press, 1989).
5 We will examine Heidegger's grounds for this claim in greater detail later in this commentary; see especially Chapter 7.
6 I will have more to say about this issue in Chapters 5 and 7.
7 For a more detailed discussion, see Chapter 3.
8 Trans. R. Taft (Bloomington, Ind.: Indiana University Press, 1990).
9 See note 2.
10 See the introductory remarks of the editor of *The Basic Problems of Phenomenology*.

1

THE HUMAN WORLD: SCEPTICISM, COGNITION AND AGENCY

(Being and Time, §§9–24)

The first division of *Being and Time* presents a preparatory fundamental analysis of Dasein. It is fundamental in so far as Heidegger's concern is ontological, or more precisely existential. He does not aim to list all of Dasein's possible existentiell modes, or to analyse any one of them, or to rely upon assumptions about human nature that have hitherto guided anthropologists, psychologists or philosophers. Instead, he offers a critical evaluation of those assumptions by developing an existential analytic of Dasein that truly allows Dasein's Being to show itself in itself and for itself. However, this fundamental analytic is also preparatory: its conclusions will not provide the terminus of his investigation, but rather a starting point from which it can be deepened, revealing the fundamental relationship between the Being of Dasein and temporality. In this sense, the first division prepares the way for the second.

The overall structure of this first division is reasonably perspicuous. An account of Dasein's average everydayness is used to demonstrate that the Being of Dasein is Being-in-the-world, which is an essentially unitary or holistic phenomenon. Heidegger thereby contests the Cartesian understanding of the human way of being as essentially compound, a synthesis of categorially distinct elements (i.e. of mind and body) in a purely material world. Nonetheless, the hyphenated elements of Being-in-the-world are relatively autonomous; so Heidegger provides separate analyses of the notion of 'world', then of the being who inhabits that world with others of its kind, and finally of the element of 'Being-in' itself. He concludes by revealing that the Being of Dasein as Being-in-the-world is founded upon and unified by what he calls 'care'. This chapter will focus upon the critique of Descartes that follows from Heidegger's analysis of the worldhood of the world; Chapter 2 will examine Dasein's relations with others and with its own affective and cognitive states; and Chapter 3 will elucidate the conceptions of language, reality and truth that follow from this conception of human existence as essentially conditioned by its world and by those with whom it occupies that world. Our discussion of Division One as a whole will conclude by elucidating the notion that Dasein's Being is essentially care (Chapter 4).

Two assumptions about the distinctive character of Dasein orient this analysis from the outset – assumptions which Heidegger initially presents simply as intuitively plausible, but later tries to elaborate more satisfactorily. The first (already introduced) is that Dasein's Being is an issue for it. The continuance of its life, and the form that life takes, confront it as questions to which it must find answers that it then lives out – or fails to. The second is this: 'that Being which is an issue for this entity in its very Being, is in each case mine' (BT, 9: 67). In part, this merely draws out one implication of the first assumption; for any entity that chooses to live in a particular way makes that existential possibility its own – that way to be becomes its way to be, that possibility becomes its own existentiell actuality. This is why Heidegger glosses his talk of Dasein's 'mineness' by saying that one must use personal pronouns when addressing it. It is his way of capturing the sense in which beings of this type are persons, but without employing such prejudicial philosophical terms as 'consciousness', 'spirit', or 'soul'; he thereby asserts that they have, if not individuality, then at least the potential for it.

These two characteristics sharply distinguish Dasein from material objects and most animals. As I emphasized earlier, tables and chairs cannot relate themselves to their own Being, not even as a matter of indifference. They have properties, some of which (what Heidegger will term their 'categories') go to make up their essence, but Dasein has – or rather is – possibilities; in so far as it has an essence, it consists in existence (whose distinguishing marks Heidegger labels 'existentialia'). But this means that human lives, unlike those of other creatures, are capable of manifesting individuality. Birds and rabbits live out their lives in ways determined by imperatives and behaviour patterns deriving from their species-identity; they instantiate their species. However, entities whose Being is in each case mine can allow *what* they are to be informed by, or infused with, *who* they are (or can fail to do so):

> [B]ecause Dasein is in each case essentially its own possibility, it can, in its very Being, 'choose' itself and win itself; it can also lose itself and never win itself; or only 'seem' to do so. But only insofar as it is essentially something which can be *authentic* – that is, something of its own – can it have lost itself and not yet won itself. As modes of Being, *authenticity* and *inauthenticity* . . . are both grounded in the fact that any Dasein whatsoever is characterized by mineness.
>
> (BT, 9: 68)

Since tables and rabbits do not, in the relevant sense, exist, they cannot be said to exist authentically or inauthentically; but since entities with the Being of Dasein do exist, they can do so either authentically or inauthentically. Inauthentic existence is not a diminution of Being; it is no less real than authentic existence. Nor is Heidegger's talk of (in)authenticity intended to embody any sort of value-judgement; it simply connotes one more distinguishing characteristic of any entity whose Being is an issue for it.

Nevertheless, this particular characteristic of Dasein motivates two other aspects of Heidegger's procedures in this part of his book. The first is the initial focus of his analysis. As we saw earlier, in order to minimize the prejudicial effects of culturally sedimented human self-understandings, he intends to orient his existential analytic around an account of Dasein in its most common, average everydayness – an essentially undifferentiated state, in which no definite existentiell mode has typically been made concrete. However, as one mode of Dasein's existence, average everydayness must also be subject to evaluation in terms of authenticity; and, according to Heidegger, it is in fact inauthentic. Although it can, therefore, perfectly legitimately be analysed in order to reveal Dasein's basic existential structures, it must not be thought of as somehow more authentic or genuine than the existentiell states typically focused upon by philosophers – states appropriate to theoretical cognition or scientific endeavour, for example.

The second thing worth noting here is Heidegger's observation that, despite the distinctiveness of Dasein's mode of Being, it is constantly interpreted in ways that fail to acknowledge it; in particular, the ontological structures appropriate to the Being of substances and physical objects are projected upon the Being of Dasein. We tend to understand Dasein in terms of what-being, as if it were possessed of an essence from which its characteristics flow in the way that a rock's properties flow from its underlying nature; we interpret ourselves as just one more entity among all the entities we encounter. Heidegger's analysis of Dasein as Being-in-the-world reveals the misconceptions underlying this interpretation; but its very prevalence, the fact that a misunderstanding of its own Being is so commonly held by the being to whom an understanding of its own Being properly and uniquely belongs, requires explanation. And his claim that authenticity is an *existentiale* of Dasein (i.e. that it is one of its existentialia) helps to provide it. For, if Dasein's average everyday state is inauthentic, then the self-understanding it embodies will be equally inauthentic; indeed, one of the distinguishing marks of Dasein's being in such a state will be its failure to grasp that which ought to be closest to it, to be most fully its own. And since philosophical enquiry is itself something that ordinary human beings do, an aspect of practical activity in human culture, the conceptions of human nature that emerge from it are likely to be similarly inauthentic.

This diagnostic move does not completely solve Heidegger's problem; for any entity capable of inauthentic existence must also be capable of authentic existence, so we still need to know why we typically end up in the former rather than the latter state – whether in philosophy or everyday life. Nonetheless, recognizing the possibility of inauthenticity at least makes it intelligible that beings, to whom an understanding of their own Being belongs, might enact their everyday existence within an inauthentic self-understanding, and proclaim that understanding as the epitome of philosophical wisdom.

THE CARTESIAN CRITIQUE (§§12–13)

The question of the human relationship with the external world has been central to Western philosophy at least since Descartes; and standard modern answers to it have shared one vital feature. Descartes dramatizes the issue by depicting himself seated before a fire and contemplating a ball of wax; when searching for the experiential roots of causation, Hume imagines himself as a spectator of a billiards game; and Kant's disagreement with Hume's analysis leads him to portray himself watching a ship move downriver. In other words, all three explore the nature of human contact with the world from the viewpoint of a detached observer of that world, rather than as an actor within it. Descartes does talk of moving his ball of wax nearer to the fire, but his practical engagement with it goes no further; Hume does not imagine himself playing billiards; and Kant never thinks to occupy the perspective of one of those sailing the ship. *Being and Time* shifts the focus of the epistemological tradition away from this conception of the human being as an unmoving point of view upon the world. Heidegger's protagonists are actors rather than spectators, and his narratives suggest that exclusive reliance upon the image of the spectator has seriously distorted philosophers' characterizations of human existence in the world.

Of course, no traditional philosopher would deny that human life is lived within a world of physical objects. If, however, these objects are imagined primarily as objects of vision, then that world is imagined primarily as a spectacle – a series of tableaux or a play staged before us; and the world of a play is one from which its audience is essentially excluded – they may look in on the world of the characters, but they do not participate in or inhabit it. Such a picture has deep attractions. A world that one does not inhabit is a world in which one is not essentially implicated and by which one is not essentially constrained; it is no accident that this spectator model attributes to the human perspective on the world the freedom and transcendence traditionally attributed to that of God. But there are also drawbacks: for the model also makes it seem that the basic human relation with objects is one of mere spatial contiguity, that persons and objects are juxtaposed with one another just as one object might be juxtaposed with another. As Heidegger puts it, it will be as if human beings are 'in' the world in just the way that a quantity of water is in a glass; and this distorts matters in two vital respects.

First, it makes this inhabitation seem like a contingent or secondary fact about human existence, rather than something which is of its essence; the water in a glass might be poured out of it without affecting its watery nature, but the idea of a human life that is not lived 'in' the world is not so easy to comprehend. Astronauts travelling beyond our planet would not thereby divest themselves of a world in the sense that interests Heidegger. Even Christian doctrines which posit a continuing personal life after our departure from the world of space and time conceive of it as involving the possession of a (resurrected) body and the inhabitation of another (heavenly) world – an environment within which they might live, move and otherwise enact their transfigured being. Heidegger's use of the term 'Dasein', with its literal meaning of 'there-being' or 'being-there', to denote the human way of being emphasizes that human existence is *essentially* Being-in-the-world; in effect, it affirms an internal relation between 'human being' and 'world'. If two concepts are internally related, then a complete grasp of the meaning of either requires grasping its connection with the other, although the two concepts are not thereby conflated. For example, pain is not reducible to pain-behaviour, but no one could grasp the meaning of the concept of pain without a grasp of what counts as behaviour expressive of pain. Heidegger's view is that the human way of being is similarly incomprehensible in isolation from a grasp of the world in which it 'is'.

The second problem with the 'spatial contiguity' model of the relation between human beings and their world is that it obliterates its distinctive nature – the proper significance of the 'in' in 'Being in-the-world'. For Heidegger, a human being confronting an object is not like one physical object positioned alongside another. A table might touch a wall, in the sense that there may be zero space between the two entities, but it cannot encounter the wall as a wall – the wall is not an item in the table's world. Only Dasein, the being to whom an understanding of Being belongs, can touch a wall in the sense that it can grasp it as such.

The ambiguity of this last phrase is instructive. Heidegger is not suggesting that philosophers such as Descartes ignored the comprehending nature of human relations to objects – after all, Descartes holds up his ball of wax precisely in order to demonstrate that human reason can penetrate to the essence of reality. But human beings can attain not only a mental or theoretical grip on objects, but also a physical or practical one – they can literally grasp them. The things Dasein encounters are usable, employable in the pursuit of its purposes: in Heidegger's terms, they are not just present-at-hand, the object of theoretical contemplation, but handy or ready-to-hand. That is the way in which Dasein encounters them when it looks after something or makes use of it, accomplishes something or leaves something undone, renounces something or takes a rest. Dasein not only comprehends the objects in its world, but also concerns itself with them (or fails to); and Heidegger feels that philosophers not only tend to pass over this phenomenon but are also unable to account for its possibility.

A Cartesian philosopher might respond to Heidegger's charge by arguing that, although she may not have paid much attention to practical interactions with the world, she can perfectly well account for readiness-to-hand on the basis of her understanding of presence-at-hand. True, Descartes' ball of wax lies on his palm, detached from any immediate practical task and from the complex array of other objects and other persons within which such tasks are pursued. The features which make it so handy for sealing letters and making candles appear as its present-at-hand characteristics, the focus of the philosopher's speculative gaze. But that gaze reveals the properties which account for its handiness for letter-writers and churchwardens; and the practical contexts within which it is so employed can be understood as compounded from a complex array of similarly present-at-hand objects and their properties, together with a story about how values and meanings are projected upon the natural world by the human mind. Such an account would demonstrate that presence-at-hand is logically and metaphysically prior to readiness-to-hand; and if it *is* explanatorily the more fundamental concept, philosophers *should* be concentrating their attention upon it.

A more detailed account of how such a strategy might work will emerge later. It is important, however, to be clear in advance about what Heidegger is and is not claiming against its proponents. He does not argue that the primacy such philosophers accord to theoretical cognition and presence-at-hand should instead be accorded to practical activity and handiness – as if building a chair were more imbued with the Being of Dasein than sitting in it to contemplate a ball of wax. Readiness-to-hand is not metaphysically prior to presence-at-hand. He *does* claim that focusing exclusively on theoretical contemplation tends to obscure certain ontologically significant aspects of that mode of activity which stand out more clearly in other sorts of case, *and which underpin both*. For, if we concentrate on cases where an immobile subject contemplates an isolated object, then our reflections upon it are likely to be significantly skewed. First, in a situation in which the human capacity for agency is idling and our understanding is preoccupied with categories appropriate to the Being of the object before us, we will tend to interpret our own nature in the terms that are readiest-to-hand – as that of one present-at-hand entity next to another. And, second, we will tend to see the relationship between these two isolated entities as itself isolated, as prior to or separable from other elements in the broader context from which we have in theory detached it, but within which that theoretical activity (just like any other activity) must in reality occur. In other words, certain features intrinsic to theoretical cognition encourage us to misinterpret its true nature, to overlook the fact that it is a species of activity, a modified form of practical engagement with the world, and so only possible (as are other, more obviously practical activities) for environed beings, beings whose Being is Being-in-the-world. But, by overlooking our worldliness, we overlook something ontologically central to *any* form of human activity, theoretical or otherwise; and, if this notion of 'world' grounds the possibility of theoretically cognizing present-at-hand objects, it cannot conceivably be explained as a construct from an array of purely present-at-hand properties and a sequence of value-projections. What is ontologically unsound is thus not theoretical cognition or presence-at-hand as such, but rather the (mis)interpretations of them – and the consequent (mis)interpretations of non-theoretical modes of activity – that have hitherto prevailed in philosophy. The true ontological importance of readiness-to-hand is that a careful analysis of it can perspicuously reveal the crucial element missing from those (mis)interpretations – the phenomenon of 'the world'.

Heidegger's discussion of Being-in-the-world therefore has a complex structure. First, he must show that practical encounters with ready-to-hand objects are only comprehensible as modes of Being-in-the-world – thus revealing the fundamental role of the hitherto unnoticed phenomenon of 'the world'. Second, he must show that theoretical encounters with present-to-hand objects are also comprehensible as a mode of Being-in-the-world – thus demonstrating that the species of human activity seemingly most suited to a Cartesian analysis can be accommodated in his own approach. And, third, he must show that a Cartesian account of readiness-to-hand is not possible – thus demonstrating that the phenomenon of 'the world' is not comprehensible as a construct from present-at-hand entities and their properties, but must be taken as ontologically primary. In the sections under consideration, Heidegger outlines his attack under the second and third headings – indicating how a phenomenological account can, and why a Cartesian account cannot, make sense of a purely cognitive relationship with entities.

He begins by pointing out that our dealings with the world typically absorb or fascinate us; our tasks, and so the various entities we employ in carrying them out, preoccupy us. Theoretical cognition of entities as present-at-hand should therefore be understood as a modification of such concern, as an emergence from this familiar absorption into a very different sort of attitude:

> If knowing is to be possible as a way of determining the nature of the present-at-hand by observing it, then there must first be a *deficiency* in our having-to-do with the world concernfully. When concern holds back from any kind of producing, manipulating and the like, it puts itself into what is now the sole remaining mode of Being-in, the mode of just tarrying-alongside. In this kind of '*dwelling*' as a holding-oneself-back from any manipulation or utilization, the *perception* of the present-at-hand is consummated.
>
> (BT, 13: 88–9)

To call 'knowing' a deficient mode of Being-in-the-world does not amount to accusing it of being less real or authentic. It implies only that it – like neglecting or taking a rest from a task – can usefully be contrasted with other sorts of activity that involve making use of objects to get something done. Only in so far as it involves holding back from interaction with objects is it 'deficient': in all other senses (and necessarily so, since it is a mode of Being-in-the-world), it is itself a fully-fledged, perfectly legitimate and potentially important way of engaging with objects. Properly understood, knowing – whether this amounts to staring at a malfunctioning tool or analysing a substance in a laboratory – is an activity carried out in a particular context, for reasons that derive from (and with results that are, however indirectly, of significance for) other human activities in other practical contexts. In short, knowing is simply one specific mode of worldly human activity, and so one node in the complex web of such activities that make up a culture and a society.

If, however, it is not properly understood, if we conceptualize it as an isolated relation between present-at-hand subject and present-at-hand object, then we face the challenge of scepticism without any way of accommodating it. For then knowledge must be conceived of as a property or possession of one or the other entity. Since it is clearly not a property of the object known, and not an external characteristic of the knowing subject, it must be an internal characteristic – an aspect of its subjectivity. In this way, the 'closet of consciousness' myth is born, and the question inevitably arises: how can the knowing subject ever emerge from its inner sanctum into the external, public realm whose entities with their properties are the supposed object of its 'knowledge'? How can such a subject ever check the supposed correspondence between its idea of an object and the object itself, when its every foray into the material realm can result only in more ideas with which to furnish its closet? How, indeed, can it ever be sure that there is an object corresponding to its ideas? As Hume famously discovered, no such demonstration is possible; and, when the very concept of an object begins to crumble, it takes with it the companion concept of an external realm, the world within which we claim to encounter objects with a life independent of their being observed by us.

Heidegger's claim (a claim that the history of philosophical attempts to refute scepticism seems to bear out) is that no answer to these sceptical challenges is possible if the subject–object relationship is understood as the being-together of two present-at-hand entities. If, however, knowing is understood as a mode of Being-in-the-world, the challenge is nullified. For 'if I "merely" know about some way in which the Being of entities is interconnected . . . I am no less alongside the entities outside in the world than when I *originally* grasp them' (BT, 13: 89–90). In short, an analysis of Dasein as essentially Being-in-the-world deprives the sceptic of any possibility of intelligibly formulating her question, whereas a Cartesian analysis deprives us of any possibility of intelligibly answering it.

This may seem like a transparent attempt to beg the question against the sceptic by dismissing the Cartesian model because it fails to refute scepticism, and then helping oneself to the very concepts that scepticism places under suspicion; but it is not. For, remember, the Cartesian investigation is meant to provide an ontologically adequate account of knowing; but, if the terms of that account make scepticism irrefutable, then they exclude the possibility of knowledge – and thereby annihilate the very phenomenon they were intended to explain. In other words, the irrefutability of scepticism in Cartesian terms constitutes a devastating internal obstacle to the

Cartesian model of the human relationship to the world. It is unable to characterize coherently the very mode of human engagement with objects that it takes to be the logical and metaphysical foundation of all our interactions with the world. And, of course, Heidegger's diagnosis locates the root of this inability in a more fundamental weakness in the Cartesian model – its failure to take account of the phenomenon of the world. For its initial interpretation of human knowledge as an isolated relation between two present-at-hand entities entirely omits that phenomenon; and the consequent irrefutability of scepticism is, in effect, a demonstration that it is not possible to arrive at a viable concept of the world if one begins from that starting point – a demonstration that the concept of the world cannot be *constructed*. One must therefore either reconcile oneself to the loss of the concept altogether, or recognize that any account of the human way of being must make use of it from the outset.

The Cartesian can, of course, protest that, whatever the lessons of the history of philosophy, it *is* possible to refute the sceptical challenge from within the Cartesian perspective and construct a viable concept of the world. And, to be sure, Heidegger cannot rely upon past failure as a guarantee of future failure. Nevertheless, the ball is very much in the Cartesian's court; and, as we delve further into Heidegger's own account of Dasein as Being-in-the-world, and gain a clearer understanding of exactly what the phenomenon of the world really is, we will discover further powerful reasons for doubting that she will be able to make good her claim.

THE WORLDHOOD OF THE WORLD (§§14–24)

According to Heidegger, the notion of 'world' can be used in at least four different ways:

1   As an ontical concept, signifying the totality of entities that can be present-at-hand within the world.
2   As an ontological term, denoting the Being of such present-at-hand entities – that without which they would not be beings of that type.

3   In another ontic sense, standing for that wherein a given Dasein might be said to exist – its domestic or working environment, for example.

4   In a corresponding ontological (or, rather, existential) sense, applying to the worldhood of the world – to that which makes possible any and every world of the third type.

Heidegger uses the term exclusively in its third sense, although his ultimate goal is to grasp that to which the term applies in its fourth sense. Consequently, the adjective 'worldly' and its cognates are properly applicable only to the human kind of Being, with physical objects or other entities described as 'belonging to the world' or 'within-the-world'. Thus, although the world must be such as to accommodate the entities encountered within it, it cannot be understood in the terms appropriate to them. The world in this third sense is one aspect of Dasein's Being, and so must be understood existentially rather than categorially (to use the Heideggerian terminology we defined in the third section of the Introduction).

Accordingly, to get the phenomenon of the world properly into view, we must locate a type of human interaction with entities that casts light on its own environment. Since certain features of theoretical, purely cognitive relations to objects tend to conceal its worldly background, Heidegger focuses instead upon a more ubiquitous and non-deficient form of human activity – that in which we make use of things, encountering them not as objects of the speculative gaze but as equipment, or more loosely as gear or stuff (as in 'cricket gear' or 'gardening stuff'). In such practical dealings with objects, they appear as ready-to-hand rather than present-at-hand; and this is where Heidegger's famous hammer makes its appearance:

> [H]ammering does not simply have knowledge about the hammer's character as equipment, but it has appropriated this equipment in a way that could not possibly be more suitable. . . . [T]he less we just stare at the hammer-Thing, and the more we seize hold of it and use it, the more primordial does our relationship to it become, and the more unveiledly is it encountered as that which it is – as

> equipment. The hammering itself uncovers the specific 'manipula-
> bility' of the hammer. The kind of Being which equipment possesses
> – in which it manifests itself in its own right – we call *readiness-
> to-hand*.

> (BT, 15: 98)

Descartes' ball of wax lies on his palm, the qualities that make it handy for sealing letters and making candles manifest as occurrent properties. But Heidegger's hammer is caught up amid a carpenter's labours, one item in a toolbox or workshop, something deployed within and employed to alter the human environment; its properties of weight and strength subserve the final product, the goal of the endeavour.

Thus, the notion of readiness-to-hand brings with it a fairly complex conceptual background that is not so evident when objects are grasped in terms of presence-at-hand, and that Heidegger aims to elucidate – handicapped as always by the fact that philosophers have hitherto ignored it, and so constructed no handy, widely accepted terminology for it. He first points out that the idea of a single piece of equipment makes no sense. Nothing could function as a tool in the absence of what he calls an 'equipmental totality' within which it finds a place – a pen exists as a pen only in relation to ink, paper, writing desks, table and so on. Second, the utility of a tool presupposes something *for which* it is usable, an end product – a pen is an implement for writing letters, a hammer for making furniture. This directedness is the 'towards-which' of equipment. Third, such work presupposes the availability of raw material; the hammer can be used to make furniture only if there is wood and metal upon which to work and from which the hammer itself can be made – that 'whereof' it is constituted. And, fourth, the end product will have recipients, people who will make use of it, and so whose needs and interests will shape the labour of the person producing the work – whether that labour is part of craft-based, highly individualized modes of production or highly industrialized ones. This is the most obvious point at which what Heidegger calls the 'public world' invades that of the workshop; here, it becomes clear that the working environment participates in a larger social world.

A piece of equipment is thus necessarily something 'in-order-to': its readiness-to-hand is constituted by the multiplicity of reference- or assignment-relations which define its place within a totality of equipment and the practices of its employment. In this sense, any single ready-to-hand object, however isolated or self-contained it may seem, is encountered within a world of work. Even in a working environment, however, this equipmental totality tends to be overlooked. For anyone concentrating on the task at hand will be focusing her attention primarily on the goal of her labours, the correctness of the final product, and the tools she is employing to achieve this will of course be caught up in the production process, rendered invisible by their very handiness. Paradoxically enough, objects become visible as ready-to-hand primarily when they become unhandy in various ways, of which Heidegger mentions three. If a tool is damaged, then it becomes *conspicuous* as something unusable; if it is absent from its accustomed place in the rack, it *obtrudes* itself on our attention as something that is not even to hand; and, if we encounter obstacles in our work, things that might have helped us in our task but which instead hinder it, they appear as *obstinately* unready-to-hand – something to be manhandled out of the way.

In all three cases the ordinary handiness of equipment becomes unreadiness-to-hand, and then presence-at-hand, as our attempts at repair or circumvention focus more exclusively on the occurrent properties with which we must now deal. Such transformations can, of course, occur in other contexts – in particular, whenever we refrain from everyday activities in order to consider the essential nature of objects – which helps explain why we then tend to reach for the category of presence-at-hand; but, in the present context, it can also bestow a certain philosophical illumination. For the unhandiness of missing or damaged objects forces us to consider with what and for what they were ready-to-hand, and so to consider the totality of assignment-relations which underpinned their handiness; and it reveals that handiness as ordinarily inconspicuous, unobtrusive and non-obstinate. In short, precisely because we cannot perform our task, the task itself, and everything that hangs together with it, is brought to our explicit awareness:

> [W]hen an assignment has been disturbed – when something is unusable for some purpose – then the assignment becomes explicit. . . . When an assignment to some particular 'towards-this' has been thus circumspectively aroused, we catch sight of the 'towards-this' itself, and along with it everything connected with the work – the whole 'workshop' – as that wherein concern always dwells. The context of equipment is lit up, not as something never seen before, but as a totality constantly sighted beforehand in circumspection. With this totality, however, the world announces itself.

(BT, 16: 105)

However, although with most pieces of equipment the world only announces itself retrospectively – when that object becomes somehow unhandy and its assignment-relations are disturbed – one type of tool is precisely designed to indicate the worldly context within which practical activity takes place: the sign. Heidegger's example is a car indicator, and, if we substitute a flashing amber light for his outmoded red arrow, his discussion becomes perfectly clear. In one sense, such a sign is simply one more piece of equipment, a tool whose proper functioning presupposes its place in a complex equipmental totality – one including the car, road-markings, conventions governing how to alter the direction of a car's travel without disrupting that of other cars, and so on. Only within that social or cultural context can the sudden appearance of a flashing amber light on the right rear bumper of a car signify that it intends to turn right. But that flashing light also lights up the environment within which the car is moving. When pedestrians and other drivers encounter it, they are brought to attend to the pattern of roads and pavements, crossings and traffic lights within which they are moving together with the signalling car, and to their position and intended movements within it. In short, the light indicates the present and intended orientation not only of the signalling car, but also of those to whom its driver is signalling; it provides a focal point around which a traveller's awareness of a manifold of equipment in the environment through which she is moving can crystallize. Heidegger puts it as follows:

A sign is . . . an item of equipment which explicitly raises a totality of equipment into our circumspection so that together with it the worldly character of the ready-to-hand announces itself.

(BT, 17: 110)

And what the world announces itself *as* is clearly neither something present-at-hand nor something ready-to-hand. For it is not itself an entity, but rather a web of socially or culturally constituted assignments within which entities can appear as the particular types of object that they are, and which must therefore always be laid out ('disclosed', as Heidegger phrases it) in advance of any particular encounter with an object. Growing up in, or otherwise coming to inhabit, a specific culture involves acquiring a practical grasp of the widely ramifying web of concepts, roles, functions and functional interrelations within which that culture's inhabitants interact with the objects in their environment. Learning to drive a car or to make furniture is a matter of assimilating that network, within which alone specific entities can appear as the entities that they are – as steering wheel, gearstick and kerb, or as tool, handle or chair. This totality makes up what Heidegger means by the world; and precisely because it is not itself an object, it is not typically an object of circumspective concern, even when it emerges from its normal inconspicuousness in ordinary practical activity. In general, it can only be glimpsed ontically in the essentially indirect manner we have just outlined. But Heidegger's concern is ontological rather than ontic; he wants to utilize such experiences as a means of access to that which underpins and makes possible the now conspicuous web of assignment-relations, to get a secure grasp on the essential nature – the worldhood – of the world.

Any piece of equipment is essentially something 'in-order-to': it is encountered as part of a manifold of equipment deployed in the service of a particular task, and so as something essentially serviceable and involved. But the widely ramifying system of reference-relations which go to make up this serviceability has a terminus:

With the 'towards-which' of serviceability there can again be an involvement: with this thing, for instance, which is ready-to-hand and

> which we accordingly call a 'hammer', there is an involvement in
> hammering; with hammering there is an involvement in making
> something fast; with making something fast, there is an involvement
> in protection against bad weather; and this protection 'is' for the
> sake of providing shelter for Dasein – that is to say, for the sake of
> a possibility of Dasein's Being.
>
> (BT, 18: 116)

Any given ready-to-hand entity is always already involved in an
(actual or potential) task which may itself be nested in other, larger
tasks; but such totalities of involvement are always ultimately
grounded in a reference-relation in which there is no further
involvement – a 'for-the-sake-of-which' that pertains to the Being
of Dasein. The handiness of a hammer is ultimately for the sake
of sheltering Dasein; the handiness of a pen is ultimately for the
sake of communicating with others. In other words, the modes of
practical activity within which entities are primarily encountered
are by their nature contributors to Dasein's modes of existence in
the world – to specific existentiell possibilities. In this sense, the
ontological structures of worldhood are and must be existentially
understood. The world is a facet of the Being of Dasein; Dasein's
Being is Being-in-the-world.

In this way, Heidegger's detailed phenomenological analysis of
Dasein as Being-in-the-world dovetails perfectly with his initial
characterization of Dasein as the being whose Being is an issue for
it; each implies the other. For, if distinctively human being is not
only life but activity, then Dasein always faces the question of which
possible mode of existence it should enact; and answering that question necessarily involves executing its intentions in practical activity.
But this in turn presupposes that Dasein exists in a world – that it
encounters a manifold of material objects as a field for such practical activity. If, then, Dasein's practical relation to its own existence
is essential to its Being, its practical relation to the world it inhabits
must also be essential Encountering objects as ready-to-hand (and
so as referred to a particular possibility of Dasein's Being) is the
fundamental ground of Dasein's Being-in-the-world.

This notion of 'world' is, of course, not at all familiar to those acquainted with the Western philosophical tradition – as Heidegger emphasizes when he contrasts his phenomenological understanding of space with the Cartesian alternative. For Descartes, space is essentially mathematicized: spatial location is fixed by imposing an objective system of coordinates upon the world and assigning a sequence of numbers to each and every item in it, and Dasein's progress through this fixed array of present-at-hand items is a matter of measuring off stretches of a space that is itself present-at-hand. On Heidegger's view, however, Dasein most fundamentally understands its spatial relations with objects as a matter of near and far, close and distant; and these in turn are understood in relation to its practical purposes. The spectacles on my nose are further away from me than the picture on the wall that I use them to examine, and the friend I see across the road is nearer to me than the pavement under my feet; my friend would not have been any closer to me if she had appeared at my side, and moving right up to the picture would in fact distance it from me. Closeness and distance in this sense are a matter of handiness and unhandiness; the spatial disposition of the manifold of objects populating my environment is determined by their serviceability for my current activities. In Heidegger's terminology, Cartesian space is an abstraction from our understanding of space as a region or set of regions, an interlinked totality of places and objects that belong to an equipmental totality and an environing work-world. Objects are in the first instance handy or unhandy, and it is their significance in that respect – rather than a pure coordinate system – that most fundamentally places them in relation to one another and to Dasein. Space and spatiality are thus neither in the subject nor in the world, but rather disclosed by Dasein in its disclosure of the world; Dasein exists spatially, it is spatial.

On the basis of this account of Dasein as Being-in-the-world, and of the worldhood of that world, Heidegger regards the logical or metaphysical priority given to presence-at-hand over readiness-to-hand in the philosophical tradition as getting things precisely the wrong way around. For him, encountering objects as present-at-hand is a mode of holding back from dealings with objects, a species of provisional and relative decontextualization, in which one is no

longer absorbed in a task to which those objects and their properties are more or less handy means. Similarly, encountering Nature
– the substances, stuffs and species of the natural world – is understood as primarily involving a task-based encounter with natural
resources which appear as the source of useful materials rather
than as something that stirs and enthrals us through its own power
and beauty, and which might then become the object of scientific
speculation. As this last example makes clear, however, recontextualization is as fundamental to Heidegger's analysis here as decontextualization. For, since such encounters with entities are legitimate
modes of Dasein's existence, and since Dasein is necessarily Being-
in-the-world, they too must be understood as essentially worldly
phenomena. Concentrating upon them may lead us to overlook the
worldly character of our existence, but that does not mean that they
are really unworldly, or any less reliant upon a (modified) totality
of assignment-relations.

Accordingly, in addition to the argument from scepticism that we
examined earlier, Heidegger has at least two main lines of attack
against those who would assign logical and metaphysical priority
to presence-at-hand, claiming that readiness-to-hand can be understood as a construct from – and so as reducible to – presence-at-hand.
First, he could argue that, in so far as encountering objects as
present-at-hand is itself a form of worldly engagement with them,
such a reductive analysis would presuppose what it was claiming to
account for. Any such analysis of readiness-to-hand requires an
account of the worldhood of the world, but any such account which
begins from the conceptual resources supplied by present-at-hand
encounters with objects would already be presupposing the phenomenon of the world. It seems evident that an understanding of a
particular landscape in terms of the resources it provides for carpen-
ters or millers is no less dependent upon a particular, culturally
determined way of conceptualizing its elements, its form and their
relation to human perception and human life, than is an understanding of it in terms of its natural beauty. But precisely analogous
points can be made about the various ways in which one can
encounter objects as present-at-hand. A carpenter who studies the
occurrent properties of a hammer with a view to repairing it does

so against the background of a particular set of assignment-relations to which she wishes to return it, and which accordingly informs the direction of her gaze and efforts. Even the scientist whose goal in studying the hammer is to comprehend its molecular structure can do so only within the complex web of equipment, resources, theory and cultural understanding (and the corresponding totality of assignment-relations) within which anything recognizable as a chemico-physical analysis of matter could even be conceived, let alone executed.[1] And when someone – perhaps a philosopher – achieves a state of genuinely disinterested attention to the objects in front of her, simply staring at them, the very disinterest she evinces is itself only possible for a being capable of being interested. As Heidegger would put it, she can tarry alongside entities only because she can also have dealings with them, so even holding back from manipulation does not occur entirely outside the ambit of worldliness. In short, even when decontextualizing really means just that – even when no recontextualization is implicitly presupposed – it cannot be understood except as a deficient mode of Being-in-the-world; so encounters with present-at-hand entities cannot intelligibly be regarded as a jumping-off point from which a conception of worldhood might be constructed.

Heidegger's second line of argument amounts to the claim that the species of worldly understanding drawn upon in encounters with objects as ready-to-hand simply could not be reduced to the species of understanding that is manifest in theoretical cognition of occurrent entities. The worldhood of the world is not comprehensible in the terms developed by speculative reason for the comprehension of present-at-hand objects and their properties. This argument is, in fact, fairly well buried in Heidegger's text: and, even when it comes to the surface, it is formulated extremely cautiously:

> The context of assignments or references, which, as significance, is constitutive for worldhood, can be taken formally in the sense of a system of Relations. But one must note that in such formalizations the phenomena get levelled off so much that their real phenomenal content may be lost, especially in the case of such 'simple' relationships as those which lurk in significance. The phenomenal content

of these 'Relations' and 'Relata' – the 'in-order-to', the 'for-the-sake-of' and the 'with-which' of an involvement – is such that they resist any sort of mathematical functionalization.

(BT, 18: 121–2)

In fact, however, as certain influential interpreters of Heidegger have stressed (perhaps most famously, Hubert Dreyfus[2]), the basis of Heidegger's argument here licenses the far stronger conclusion that the worldhood of the world is simply not analysable in such terms.

The argument rests on two tightly interlinked points: the indefinability of context, and the difference between knowing how and knowing that. First, the point about context. The capacity to encounter a pen as a handy writing implement or a hammer as a carpentry tool depends upon a capacity to grasp its role in a complex web of interrelated equipment in certain sorts of context; but spelling out its relations with such totalities is far from simple. A hammer is not just something for driving nails into surfaces: anyone who understands its nature as a tool also knows which kinds of surface are appropriate for receiving nails, the variety of substances from which a usable hammer can be made, the indefinite number of other tasks that a hammer can be used to perform (securing wedges, loosening joints, propping open windows, repelling intruders, playing games of 'toss-the-hammer' and so on), of other objects that might be used instead of a damaged hammer or adapted so as to be usable in these ways – the list goes on. Knowing what it is for something to be a hammer is, among other things, knowing all this; and knowing all this is an inherently open-ended capacity – one which cannot be exhaustively captured by a finite list of precise rules. Our practical activities always engage with and are developed in specific situations, but there is no obvious way of specifying a closed set of all the possible ways and contexts in which our knowledge of a hammer and its capacities might be pertinently deployed. In so far as any attempt to reduce readiness-to-hand to presence-at-hand necessarily involves reducing our understanding of an object's serviceability to a grasp of a finite set of general rules together with a precise specification of a finite set of situations in which they apply, then it is doomed from the outset.

This brings us to the second of the issues mentioned above – the difference between knowing how and knowing that. Encountering a hammer as ready-to-hand is, as we have seen, intimately related to a capacity to make use of it as the piece of equipment it is – the capacity to hammer. This is a species of practical ability, manifest in the first instance in competent action, in what we might call know-how; but theoretical cognition, as understood by the philosophical tradition, is primarily manifest in a grasp of true propositions, in what might be called knowing that (such-and-such is the case). To argue that the readiness-to-hand of a hammer can be understood as a construct from its occurrent properties together with certain facts about its relations with particular contexts of action thus amounts to arguing that know-how can be understood in terms of knowing that – as the application of knowledge of facts about the object, the situation and the person wishing to employ it in that situation. Ever since the time of Ryle's *Concept of Mind*,[3] however, this idea has been under severe pressure, since its proponents face a dilemma. For the propositional knowledge they invoke must be applied to the situations the knower faces, a process which must itself either be based on further propositional knowledge (a knowledge of rules governing the application of the theorems cognized) or entirely ungrounded. If the former option is chosen, it follows that applying the rules of application must itself be governed by application rules, and an infinite regress unfolds. If the latter is preferred, the question arises why the original practical ability cannot itself be ungrounded: if the theorems can be applied without relying upon propositional knowledge, why not the actions that the theorems were designed to explain? In short, the idea that know-how is based upon knowing that involves assigning a role to propositional knowledge which it is either impossible or unnecessary for it to perform; so the idea that the knowledge manifest in our encounters with ready-to-hand objects can be reduced to knowledge of the sort appropriate to encounters with present-at-hand objects must be either vacuous or superfluous.

Putting these two lines of argument together with the argument from scepticism suggests that Heidegger can meet the challenge posed by the Cartesian philosopher to his analysis of Dasein as

Being-in-the-world. His concept of 'world' does not illegitimately give priority to systems of value that are merely subjective projections upon an ultimately meaningless but metaphysically fundamental realm of matter; it rather constitutes the ontological underpinning of any and every mode of human engagement with objects, including the seemingly value-neutral theoretical encounters of which philosophers are generally so enamoured.

Even here, however, a worry can resurface about the strength of Heidegger's case: the worry that it is undermined by a perfectly obvious fact about material objects – namely, their materiality. For surely no object can be encountered as ready-to-hand or as present-at-hand unless it is actually *there* to be encountered and possessed of certain properties; a hammer could not be used for hammering unless it had the requisite weight, composition and shape, and it could not even be contemplated unless it was actually there before us. But, if so, if any form of human encounter with an object presupposes its material reality, must not the whole web of culturally determined assignment-relations that constitutes the world of human practical activity be conceptually or metaphysically dependent upon the material realm within which human culture emerges and without which it could not be sustained? Is it not obvious that 'the world' in the third and fourth senses of that term presupposes 'the world' in the first and second senses?

This worry should not be dismissed lightly; but it is one that Heidegger only confronts in convincing detail much later – in his reflections on truth and reality (which we will examine in Chapter 3 of this book). He does, however, attempt to assuage the worry at this point, so I will conclude this chapter by outlining his strategy. The crucial move is to distinguish the ontic and the ontological levels of analysis, and to suggest that the worry I have just articulated conflates the two. Heidegger never denies that a hammer could not be used for hammering unless it had the appropriate material properties and was actually available for use; in this sense, the materiality of any given object is needed to explain its functioning. But this is an issue on what he would call the ontic level – the level at which we concern ourselves with particular (types of) human practices and the particular (types of) objects that are involved in them, and simply take it for granted that there are such practices and that within them objects are encountered as ready-to-hand, unhandy and present-at-hand. At the ontological level, however, we put exactly those assumptions in question: we enquire into the Being of human practical activity and of material objects, asking what must be the case for there to be a human world of practical activity, and what the readiness-to-hand, unhandiness or presence-at-hand of an object really amounts to. It is to this task that Heidegger has devoted these opening sections of his book. His line of argument entails that, if we are to understand the essential nature (the Being) of any of these phenomena, then we must invoke the notion of 'world' and its ontological presuppositions. Those presuppositions are not only impossible to account for in terms of the categories appropriate to species of theoretical cognition, but must themselves be invoked to account for the ontological presuppositions of theoretical cognition itself. By overlooking or downplaying the concept of 'the world' in its third and fourth senses, therefore, philosophers have prevented themselves from understanding both the mode of human activity in which we most often engage, and also that to which they accord the highest priority; and they thereby deprive themselves of any proper understanding of the Being of Dasein.

NOTES

1   Heidegger sketches in further details of such an account of scientific endeavour in §69 of *Being and Time*, which we will discuss in Chapter 6.

2   See especially ch. 6 of his *Being-in-the-World* (Cambridge, Mass.: The MIT Press, 1991).

3   G. Ryle, *The Concept of Mind* (London: Hutchinson, 1949).

2

THE HUMAN WORLD: SOCIETY, SELFHOOD AND SELF-INTERPRETATION

(Being and Time, §§25–32)

It should already be becoming clear that Heidegger conceives of the human way of being as essentially conditioned. The Western philosophical tradition has often presupposed that the human subject can in some way transcend the material realm upon which it fixes its gaze, and so that human beings are only contingently possessed of a world; but, for Heidegger, no sense attaches to the idea of a human being existing apart from or outside a world. This does not, however, mean that human beings are somehow imprisoned in the world, forcibly subjected to the essentially alien limits of embodiment and practical interaction with nature; for those limits are not essentially alien. If no recognizably human existence is conceivable in the absence of a world, then the fact that human existence is worldly cannot be a limitation or constraint upon it; just as someone can only be imprisoned if there is a world outside her prison from which she is excluded, so a set of limits can only be thought of as limitations if there exists a possible mode of existence to which those limits do not apply. Since that is not the case here, the inherent worldliness of human existence must be thought of as an aspect of the human condition. It is a condition of human life, not a constraint upon it.

But, on Heidegger's account, human existence is not only conditioned by worldliness – or, rather, worldliness conditions human existence in ways that we have not yet examined. This chapter will examine two of them: the way in which the world is inherently social or communal, and the ways in which it conditions human affective and cognitive powers.

INDIVIDUALITY AND COMMUNITY (§§25–7)

So far, it may have seemed that Dasein's world is populated solely by physical objects or entities, what J. L. Austin called 'medium-sized dry goods'. But Heidegger emphasizes that there is at least one other class of beings that must be accommodated by any adequate analysis of that world, those with the kind of Being belonging to Dasein – in short, other people. And if we cannot understand Dasein in the terms appropriate to objects, then neither can we understand other human beings and Dasein's relations with them in that way.

But, of course, many philosophers have tried to do just that. The very title under which this set of issues is commonly known in the discipline confirms this: 'The Problem of Other Minds'. It implies that, while we can be certain of the existence of other creatures with bodies similar to our own, justifying the hypothesis that these bodies have minds attached to them is deeply problematic. Here, a dualistic understanding of human beings as mind–body couples combines with a materialist impulse to suggest that our relations with other putatively human beings are, in effect, relations with physical objects of a particular sort to which we are inclined to attribute various distinctive additional characteristics – which inevitably raises the question of our warrant for such extremely unusual attributions. And any attempts to solve this 'problem' inevitably share those presuppositions, since they will be couched in the terms in which the problem itself is posed.

The argument from analogy, for example, tells us that our justification lies in the similarities of form and behaviour between our bodies and those of other humanoid creatures. Given that we know from our own case that such behaviour is associated with mental activities of various sorts, we can reliably infer that the same is true in the case of these other entities. This is a species of inductive inference, drawing a conclusion about what is correlated with the behaviour of other bodies on the basis of our acquaintance with what is correlated with the behaviour of our own. But, of necessity, our observations relate solely to correlations between mental phenomena and our own behaviour, and so provide no basis whatever for conclusions about what (if anything) might be correlated with the behaviour of others – a correlation that it is in principle impossible for us to observe directly. It may seem that such an extrapolation is justified by observable similarities between our own bodies and behaviour and the bodies and behaviour of others, but the key issue is: *which* similarities? That the bodies and the behaviour are similar in bodily and behavioural respects is not in question. But the similarity that matters is that a mind be similarly attached to those other bodies and their behaviour; and no amount of similarity between *our* bodily form and behavioural repertoire and *theirs* can establish that. To think otherwise – to think that a correlation established between body and mind in my own case can simply be extrapolated to the case of others – is to assume that comprehending the essential nature of others is simply a matter of projecting our understanding of our own nature onto them. But it is precisely the legitimacy of such empathic projection – of regarding (one's relation to) another humanoid creature as if it were just like (one's relation to) oneself, or, in more Heideggerian language, viewing Being-towards-Others in terms of Being-towards-oneself – that is at issue.

This, I take it, is Heidegger's point in the following passage:

> The entity which is 'other' has itself the same kind of Being as Dasein. In Being with and towards Others, there is thus a relationship of Being from Dasein to Dasein. But it might be said that this relationship is already constitutive for one's own Dasein, which, in its own

right, has an understanding of Being, and thus relates itself towards Dasein. The relationship-of-Being which one has towards Others . . . then become[s] a Projection of one's own Being-towards-oneself 'into something else'. The other would be a duplicate of the Self.

But while these deliberations seem obvious enough, it is easy to see that they have little ground to stand on. The presupposition which this argument demands – that Dasein's Being towards an Other is its Being towards itself – fails to hold. As long as the legitimacy of this presupposition has not turned out to be evident, one may still be puzzled as to how Dasein's relationship to itself is thus to be disclosed to the Other as Other.

(BT, 26: 162)

Thus, the argument from analogy appears to work only if the question it is designed to answer is begged – only if it is assumed from the outset that all the other humanoid bodies I encounter are similar to mine not only physically and behaviourally but also psychophysically, i.e. that they are similarly correlated with minds. The similarity that legitimates the inductive inference thus turns out to be the similarity that it is supposed to demonstrate; the argument from analogy assumes what it sets out to prove. In this respect, a Cartesian understanding of other minds faces the same difficulty as a Cartesian understanding of the external world: in both cases, no satisfactory answer is available to the sceptical challenge that the terms of such understandings invite. Heidegger concludes that we should therefore jettison an essentially compositional understanding of other persons: the sceptic's ability to demolish our best attempts to treat that concept as a construction from more basic constituents (e.g. as resulting from the projection of the concept of a humanoid mind on to that of a humanoid body) reveals that such treatments either presuppose or eliminate what they set out to analyse. We must, rather, recognize that the concept of the Other (of other persons) is irreducible, an absolutely basic component of our understanding of the world we inhabit, and so something from which our ontological investigations must begin. To adapt Strawsonian terminology, it is the concept of other persons (and not that of other minds plus other bodies) that is logically primitive.[1] And in so far as others are primordially persons, creatures with a perspective upon the world and whose essence is existence, then their Being must be of the same kind as Dasein.

But Heidegger's point is anti-solipsistic as well as anti-dualist. It is not just that the concept of another person must be understood non-compositionally (i.e. as Dasein rather than as the juxtaposition of two present-at-hand substances). That concept is also essential to any adequate ontological analysis of Dasein (i.e. the Being of Dasein is essentially Being-with-Others). After all, the Being of Dasein is Being-in-the-world, so the concepts of Dasein and world are internally related. But the structure of the world makes essential reference to other beings whose Being is like Dasein's own. So Dasein cannot be understood except as inhabiting a world it necessarily shares with beings like itself.

And just what are these essential references to Others?

> In our description of the ... work-world of the craftsman ... the outcome was that along with the equipment to be found when one is at work, those Others for whom the work is destined are 'encountered too'. If this is ready-to-hand, then there lies in the kind of Being which belongs to it (that is, in its involvement) an essential assignment or reference to possible wearers, for instance, for whom it should be cut to the figure. Similarly, when material is put to use, we encounter its producer or supplier as one who 'serves' well or badly. . . . The Others who are thus 'encountered' in a ready-to-hand, environmental context of equipment are not somehow added on in thought to some Thing which is proximally just present-at-hand; such 'Things' are encountered from out of the world in which they are ready-to-hand for Others – a world which is always mine too in advance.
>
> (BT, 26: 153–4)

This suggests three different senses in which other people are constituents of Dasein's world. First, they form one more class of being that Dasein encounters within its world. Second, what Dasein works upon is typically provided by others and what it produces is typically destined for others; in other words, the 'whereof' and the 'towards-which' of equipmental totalities relate the work-world to other people. Third, the readiness-to-hand of objects for a particular Dasein is not (and could not conceivably be) understood as their readiness-to-hand for that Dasein alone; if any object is handy for a given task, it must be handy for every Dasein capable of performing it. In this sense, readiness-to-hand is inherently intersubjective; and since a parallel argument applies to the recontextualized world of present-at-hand objects, it entails that Dasein's inherently worldly Being is essentially social.

Note that Heidegger is not claiming that Dasein cannot be alone, isolated from all human company; whether or not that is the case is a purely ontic question, to do with a particular individual in a particular time and place. The claim that the Being of Dasein is Being-with is an ontological claim; it identifies an existential characteristic of Dasein which holds regardless of whether an Other is present, and for two reasons. First, because, if it did not, the possibility of Dasein's encountering another creature of its own kind would be incomprehensible. For, if, ontologically, Dasein's Being was not Being-with, it would lack the capacity to be in another's company – just as a table can touch a wall but can never encounter it as a wall, so Dasein could never conceivably encounter another human being as such. Second, it is only because Dasein's Being is Being-with that it *can* be isolated or alone; for, just as it only makes sense to talk of Dasein encountering an object as unready-to-hand if it can also encounter it as handy, so it only makes sense to talk of Dasein as being alone if it is capable of being with Others when they are present. In other words, aloneness is a deficient mode of Dasein's Being; 'The Other can be missing only *in* and *for* a Being-with' (BT 26: 157).

The same distinction between ontic and ontological matters underpins Heidegger's further claim that, just as Dasein's basic orientation towards ready-to-hand objects is one of concern, so its orientation towards Others is one of solicitude. For, of course, 'concernful' dealings with objects can take the form of indifference, carelessness and neglect: the term captures an aspect of Dasein's ontological state, highlighting the fact that Dasein finds itself amid objects with which it must deal, and is not only compatible with, but ultimately makes possible, specific ontic states of unconcern (since it is only to a being capable of concern that one can attribute lack of concern). Similarly, talk of Dasein's Being-with-Others as solicitude is an ontological claim: it does not deny that Dasein can be and often is indifferent or hostile to the well-being of others, but rather brings out the onto-logical underpinning of all specific ontic relations to one's fellow human beings, whether they be caring or aggressive.

Heidegger sees no conflict between his claim that Dasein's Being is Being-with and his earlier characterization of Dasein's Being as in each case mine; rather, the former constitutes a further specifica-tion of the latter. That notion of 'mineness' encapsulates two main points: first, that the Being of Dasein is an issue for it (that every choice it makes about which existentiell possibilities to realize is a choice about the form that *its own* life will take), and, second, that each Dasein is an individual, a being to whom personal pronouns can be applied and to whom at least the possibility of genuine or authentic individuality belongs. To go on to claim that the Being of such a being is Being-with does not negate that prior attribution of mineness; for to say that the world is a social world is simply to say that it is a world Dasein encounters as 'our' world, and such a world is no less mine because it is also yours. Our world is both mine and yours; intersubjectivity is not the denial of subjectivity but its further specification. And this further specification deepens our under-standing of the condition under which each Dasein must develop (or fail to develop) its mineness or individuality. For, if Dasein's Being is Being-with, an essential facet of that which is an issue for Dasein is its relations to Others; the idea is that, at least in part, Dasein establishes and maintains its relation to itself in and through its relations with Others, and vice versa. The two issues are ontologi-cally inseparable; to determine the one is to determine the other.

This understanding of the relationship between subjectivity and intersubjectivity determines Heidegger's characterization of Dasein's average everyday mode of existence. For it entails that Dasein's capacity to lose or find itself as an individual always determines, and is determined by, the way in which Dasein understands and conducts its relations with Others. And the average everyday form of that understanding focuses upon one's differences (in appearance, behaviour, lifestyle and opinion) from those with whom one shares the world, regarding them as the main determinant of one's own sense of self. Our usual sense of who we are, Heidegger claims, is purely a function of our sense of how we differ from others. We understand those differences either as something to be eliminated at all costs, thus taking conformity as our aim; or (perhaps less commonly) as something that must at all costs be emphasized and developed – a strategy which only appears to avoid conformity, since our goal is then to distinguish ourselves from others rather than to distinguish ourselves in some particular, independently valuable way, and so amounts to allowing others to determine (by negation) the way we live. The dictatorship of the Others and the consequent loss of authentic individuality in what Heidegger calls 'average everyday distantiality' is therefore visible not just in those who aim to read, see and judge literature and art as everyone reads, sees and judges, but also in those whose aim is to adopt the very opposite of the common view. Cultivating uncommon pleasures, thoughts and reactions is no guarantee of existential individuality.

> Dasein, as everyday Being-with-one-another, stands in subjection to Others. It itself is not; its Being has been taken away by the Others. Dasein's everyday possibilities of Being are for the Others to dispose of as they please. These Others, moreover, are not *definite* Others. On the contrary, any Other can represent them. . . . One belongs to the Others oneself and enhances their power. The Others whom one thus designates in order to cover up the fact of one's belonging to them essentially oneself, are those who proximally and for the most part *'are there'* in everyday Being-with-one-another. The 'who' is not this one, not that one, not oneself, not some people, and not the sum of them all. The 'who' is the neuter, *the 'they'*.
>
> (BT, 27: 164)

In other words, this absence of individuality is not restricted to some definable segment of the human community; on the contrary, since it defines how human beings typically relate to their fellows, it must apply to most if not all of those Others to whom any given Dasein subjects itself. They cannot be any less vulnerable to the temptations of distantiality, and so cannot be regarded as having somehow avoided subjection to those who stand as Others to them. 'The Others' thus cannot be thought of as a group of genuinely individual human beings whose shared tastes dictate the tastes of everyone else; and neither do they constitute an intersubjective or supra-individual being, a sort of communal self. The 'they' is neither a collection of definite Others nor a single definite Other; it is not a being or set of beings to whom genuine mineness belongs, but a free-floating, impersonal construct, a sort of consensual hallucination to which each of us gives up the capacity for genuine self-relation and the leading of an authentically individual life. Consequently, if a given Dasein's thoughts and deeds are (determined by) what *they* think and do, its answerability for its life has been not so much displaced (on to others) as misplaced. It has vanished, projected on to an everyone that is no one by someone who is, without it, also no one, and leaving in its wake a comprehensively neutered world. As Heidegger puts it, 'everyone is the other and no one is himself. The "*they*", which supplies the answer to the question of the "who" of everyday Dasein, is the "*nobody*" to whom every Dasein has already surrendered itself in Being-among-one-another' (BT, 27: 165–6).

In short, the average everyday mode of Dasein is inauthentic. Its mineness takes the form of the 'they', its Self is a they-self – a mode of relating to itself and to Others in which it and they fail to find themselves and so fail to achieve genuine individuality. And this cultural critique also accounts for the prevalence of ontological misunderstandings in the philosophical tradition. For Heidegger needs to explain how a creature to whom (according to his own analysis) an understanding of Being essentially belongs can have misunderstood its own Being so systematically. But, of course, if Dasein typically loses itself in the 'they', it will understand both its world and itself in the terms that 'they' make available to it, and so will interpret its own nature in terms of the categories that lie closest to hand in popular culture and everyday life; and they will be as inauthentic as their creators. They will embody the same impulses towards levelling down, the avoidance of the unusual or the difficult, the acceptance of prevailing opinion, and so on. And since philosophical enquiry will typically be the work of those same inauthentic individuals, the philosophical tradition will contain similarly inauthentic ontological categories that are unhesitatingly accepted by its present representatives. Any attempt to retrieve an authentic ontological understanding will accordingly appear to subvert obvious and self-evident truths, to overturn common sense and violate ordinary language.

Two words of warning are in order about this notion of inauthenticity. First, such an inauthentic state is not somehow ontologically awry, as if Dasein were less real as an entity, less itself, when its Self is the they-self. On the contrary, any Being capable of finding itself must also be capable of losing itself. Second, authenticity does not require severing all ties with Others, as if genuine individuality presupposed isolation or even solipsism. Heidegger's view is rather that Dasein's Being is Being-with; in other words, just as with Dasein's worldliness, its inherently social forms of existence are not a limitation upon it but a limit – a further condition of the human way of being. So authentic Being-oneself could not involve detachment from Others; it must rather require a different form of relationship with them – a distinctive form of Being-with.

Unfortunately, Heidegger's way of stating this last point raises more questions than it answers. For he says that 'authentic Being-oneself is . . . an existentiell modification of the "they" – of the "they" as an essential *existentiale*' (BT, 27: 168). If the they-self is an essential existentiale of Dasein, it is not just a particular existentiell possibility that Dasein commonly tends to actualize, but rather a 'primordial phenomenon [which] belongs to Dasein's positive constitution' (BT, 27: 167), part of its ontological structure. But since submission to the they-self is an inherently inauthentic mode of Dasein's Being, Heidegger seems to be claiming that Dasein's Being is somehow inherently inauthentic. In other words, whereas previously he has claimed that Dasein is ontologically capable of living either authentically or inauthentically, and that which it achieves depends upon where, when and how it makes its existentiell choices, now he wants to claim that Dasein's very nature mires it in an inauthenticity of which such authenticity as it may sometimes achieve is merely an existentiell modification.

It is hard to see what sense might be attached to the idea that authenticity is an existentiell mode of an ontologically inauthentic being; how can Dasein be both authentic and inauthentic at once – authentically inauthentic? More generally, Heidegger's claim looks like a simple confusion of his own categories, a blurring of the very distinction between ontic and ontological levels of analysis to which he constantly makes reference; and his analysis in this chapter provides no support for the conclusion he wants to draw. For its focus is Dasein's average everydayness, which is an existentiell state, and so can reveal only that the Self of *everyday* Dasein is the they-self. If this licenses any ontological conclusion – a conclusion concerning structures of Dasein's Being regardless of its particular ontic state – it is that Dasein's Being is always Being-with. It certainly does not license the conclusion that that Being-with must take the inauthentic form of submission to the 'they'.

Can Heidegger's seeming waywardness here be justified, or at least accounted for? Two passages provide a clue, the first from the beginning of section 27:

> We have shown earlier how in the environment which lies closest to us, the 'public' environment already is ready-to-hand and is also a matter of concern. In utilizing means of transport and in making use of information services such as the newspaper, every Other is like the next. This Being-with-one-another dissolves one's own Dasein completely into the kind of Being of 'the Others', in such a way, indeed, that the Others, as distinguishable and explicit, vanish more and more.
>
> (BT, 27: 164)

In one sense, this passage gets us no further forward, since the phenomena it picks out (prevailing arrangements for transport and newspapers) are features of Dasein's world that one can easily imagine being altered more or less radically; there seem to be no ontological implications here. On the other hand, it plainly links the idea of one Dasein being just like the next with that of the environment that lies closest to it, which is of course the work-world – as if for Heidegger there is something inherently public or impersonal about that world, something that no more acknowledges the individuality of those who inhabit it than a public transportation system acknowledges the individuality of each of its 'customers' or a newspaper that of each of its readers. What might this something be?

The second passage appears a little earlier:

> [W]hen material is put to use, we encounter its producer or 'supplier' as one who 'serves' well or badly. When for example, we walk along the edge of a field but 'outside it', the field shows itself as belonging to such-and-such a person, and decently kept up by him. The book we have used was bought at so-and-so's shop. . . . The boat anchored at the shore is assigned in its being-in-itself to an acquaintance that undertakes voyages with it; but even if it is a 'boat which is strange to us', it is still indicative of Others.
>
> (BT, 26: 153–4)

At first, this passage seems only to emphasize the multitude of ways in which Dasein's world reveals the presence of Others; but, reading it with our problem in mind, what might strike us instead is just *how* those Others appear to Dasein. They appear as producers, suppliers, field-owners and farmers, booksellers and sailors – in short, as bearers of social roles; and they are judged in terms of how well or badly they carry out their roles. Their identity is thus given primarily by their occupation, by the tasks or functions they perform; who they are to us is a matter of what they do and how they do it. But these are defined purely impersonally, by reference to what the relevant task or office requires; given the necessary competence, which individual occupies that office is as irrelevant as are any idiosyncrasies of character and talent that have no bearing on the task at hand. In so far, then, as Others appear in our shared world primarily as functionaries, they appear not as individuals but as essentially interchangeable occupants of impersonally defined roles. Since our appearance to them must take a precisely analogous form, we must understand ourselves to be in exactly the same position.

We can see why this is an ontological rather than an ontic matter if we recall Heidegger's earlier analysis of the worldhood of the world. It constitutes a widely ramifying web of socially defined concepts, roles, functions and functional interrelations, within which alone it was possible for human beings to encounter objects. Heidegger's analysis of Dasein's Being as Being-with simply underlines the fact that human beings, no less than objects, are part of that same web; after all, their Being is Being-in-the-world. Since the environment closest to them is the work-world, the identity closest to them is their identity as workers, as people performing socially defined and culturally inherited tasks whose nature is given prior to and independently of their own individuality, and which typically will not be significantly marked by their temporary inhabitation of them. Just as the objects with which we deal must be understood primarily in relation to purposes and possibilities-of-Being embedded in cultural practices, so we must understand ourselves primarily as practitioners – as followers of the norms definitive of proper practice in any given field of endeavour. And Heidegger's point is that such norms – and so such practices – are necessarily interpersonal, and so in an important sense impersonal. It must be possible for others to occupy exactly the same role, to engage in exactly the same practice; apart from anything else, society and culture could not otherwise be reproduced across generations. But, more importantly, a practice that only one person could engage in simply could not count as a practice at all. Such a thing would be possible only if it were possible for someone to follow a rule that no one else could follow – to follow a rule privately – and as Wittgenstein has argued, that is a contradiction in terms.[2]

For Heidegger, then, since Dasein's Being is Being-in-the-world, it will always, necessarily, begin from a position in which it must relate to itself as the occupant of a role in a practice, and so must begin by understanding itself in the essentially impersonal terms that such a role provides – terms which have no essential connection with its identity as an individual, but rather define a function or set of functions that anyone might perform. Such roles do not, as it were, pick out a particular person, even if they do require particular skills or aptitudes; they specify not what you or I must do in order to occupy them, but rather what one must do – what must be done. The role-occupant thus specified is an idealization or construct, an abstract or average human being rather than anyone in particular:

it is, in other words, a species of the they-self. In this sense, and this sense alone, is the 'they' an essential *existentiale* of Dasein.

But, of course, just because such roles are defined in entirely impersonal terms, the individual who occupies them need not always relate to them purely impersonally. A social role can be a vital element in an individual's self-understanding (as a vocation, for example); but, although the role can be appropriated authentically in such ways, its essential nature does not ensure or even encourage such appropriations. Heidegger does not deny the possibility of authentic existence to beings who must begin from such a self-understanding. He simply claims that the position from which they must begin necessarily involves a self-interpretation from which they must break away if they are to achieve authentic existence, and that any such authentically individual existence, since it must be lived in the world, must be a modification rather than a transcendence of the role-centred nature of any such life. Authenticity is a matter of the way in which one relates to one's roles, not a rejection of any and all roles. In short, Dasein is never necessarily lost to itself, but it must always begin by finding itself; authenticity is always an achievement:

> The Self of everyday Dasein is the *they-self*, which we distinguish from the *authentic Self* – that is, from the Self which has been taken hold of in its own way. As they-self, the particular Dasein has been *dispersed* into the 'they', and must first find itself. . . . If Dasein discovers the world in its own way and brings it close, if it discloses to itself its own authentic Being, then this discovery of the 'world' and this disclosure of Dasein are always accomplished as a clearing-away of concealments and obscurities, as a breaking up of the disguises with which Dasein bars its own way.
>
> (BT, 27: 167)

PASSIONS AND PROJECTS (§§28–32)

After examining the notion of 'world' and the species of selfhood Dasein typically exhibits, Heidegger turns to the notion of 'Being-in' – the third and final element in the structural totality of Being-in-the-world. His aim is to deepen his earlier, introductory remarks about this third notion, going beyond their primarily anti-Cartesian animus towards a more positive ontological analysis that draws upon his accounts of worldhood and selfhood. For, of course, each element in Dasein's ontological structure is only relatively autonomous: analytical clarity is furthered by examining each with some degree of independence, but analytical accuracy demands that we recognize that they are internally related – the significance of each ultimately inseparable from that of the ontological whole they make up. With respect to 'Being-in', that means recognizing that the way in which Dasein inhabits its world reflects and determines the nature of the world thus inhabited, and in particular that it is a world in which Dasein dwells together with others just like itself – a social world.

The more particular focus of this new investigation of 'Being-in', however, involves the fact that Dasein's relation to its world, its being-there or there-being, is a comprehending one. Heidegger underlines this in a potentially misleading but nonetheless illumi- nating way by claiming that, in so far as we think of our commerce with the world as a relation between subject and objects, then Dasein is the Being of this 'between'. In other words, he recognizes that Dasein is not trapped within a mind or body from which it then attempts to reach out to objects, but is, rather, always already outside itself, dwelling amid objects in all their variety. Dasein's thoughts, feelings and actions have entities themselves (not mental represen- tations of them) as their objects, and those entities can appear not merely as environmental obstacles or as objects of desire and aver- sion, but in the full specificity of their nature, their mode of existence (e.g. as handy, unready-to-hand, occurrent, and so on), and their reality as existent things. This capacity to encounter entities *as* enti- ties is what Heidegger invokes when he talks of Dasein as the clear- ing, the being to whom and for whom entities appear as they are:

> Only for an entity which is existentially cleared in this way does that which is present-at-hand become accessible in the light or hidden in the dark. By its very nature, Dasein brings its 'there' along with it. If it lacks its 'there', it is not factically the entity which is essentially Dasein; indeed, it is not this entity at all. *Dasein is its disclosedness.*
>
> (BT, 28: 171)

In this section we shall examine Heidegger's claim that the existential constitution of Dasein's Being-in has two elements – state-of-mind and understanding – both of which constitute limits or conditions of distinctively human existence.

What Heidegger labels 'Befindlichkeit' is an essentially passive or necessitarian aspect of Dasein's disclosure of itself and its world. The standard translation of 'Befindlichkeit' as 'state-of-mind' is seriously misleading, since the latter term has a technical significance in the philosophy of mind which fails to match the range of reference of the German term. Virtually any response to the question 'How are you?' or 'How's it going?' could be denoted by 'Befindlichkeit' but not 'state-of-mind'. The latter also implies that the relevant phenomena are purely subjective states, thus repressing Heidegger's emphasis upon Dasein as Being-in-the-world. 'Frame of mind' is less inaccurate, but still retains some connotation of the mental as an inner realm. Consequently it seems best to interpret 'Befindlichkeit' as referring to Dasein's capacity to be affected by the world, to find that the entities and situations it faces *matter* to it, and in ways over which it has less than complete control.

The most familiar existentiell manifestation of this *existentiale* is the phenomenon of mood. Depression, boredom and cheerfulness, joy and fear, are affective inflections of Dasein's temperament that are typically experienced as 'given', as states into which one has been thrown – something underlined in the etymology of our language in this region. We talk, for example, of moods and emotions as 'passions', as something passive rather than active, something that we suffer rather than something we inflict – where 'suffering' signifies not pain but submission, as it does when we talk of Christ's Passion or of His suffering little children to come unto Him. More generally, our affections do not just affect others but mark our having been affected by others; we cannot, for example, love and hate where and when we will, but rather think of our affections as captured by their objects, or as making us vulnerable to others, open to suffering.

For human beings, such affections are unavoidable and their impact pervasive. They constitute a further and fundamental condition of human existence. We can, of course, sometimes overcome or alter our prevailing mood, but only if that mood allows, and only by establishing ourselves in a new one (tranquillity and determination are no less moods than depression or ecstasy); and once in their grip, moods can colour every aspect of our existence. In so doing, of course, they determine our grasp upon the world: they inflect Dasein's relation to the objects and possibilities among which it finds itself – one and all being grasped in relation to the actualized possibility-of-Being that Dasein is. In this sense, moods are disclosive: a particular mood discloses something (sometimes everything) in the world as mattering to Dasein in a particular way – as fearful, boring, cheering or hateful; and this reveals in turn that, ontologically speaking, Dasein is open to the world as something that can affect it.

It is, however, easier to accept the idea that moods disclose something about Dasein than that they reveal something about the world. Since human beings undergo moods, the claim that someone is bored or fearful might be said to record a simple fact about her. But her mood does not – it might be thought – pick out a simple fact about the world (namely, that it is, or some things within it are, boring or fearsome), for moods do not register objective features of reality but rather subjective responses to a world that is in itself essentially devoid of significance. In short, there can be no such thing as an epistemology of moods. Heidegger, however, wholeheartedly rejects any such conclusion. Since moods are an aspect of Dasein's existence, they must be an aspect of Being-in-the-world – and so must be as revelatory of the world and of Being-in as they are of Dasein. As he puts it:

> A mood is not related to the psychical . . . and is not itself an *inner condition* which then reaches forth in an enigmatical way and puts its mark on things and persons. . . . It comes neither from 'outside' nor from 'inside', but arises out of Being-in-the-world, as a way of such being.
>
> (BT, 29: 176)

Heidegger reinforces this claim with a more detailed analysis of fear. Its basic structure has three elements: that in the face of which we fear, fearing itself, and that about which we fear. That in the face of which we fear is the fearful or the fearsome – something in the world which we encounter as detrimental to our well-being or safety; fearing itself is our response to that which is fearsome; and that about which we fear is of course our well-being or safety – in short, ourselves. Thus, fear has both a subjective and an objective face. On the one hand it is a human response, and one that has the existence of the person who fears as its main concern. This is because Dasein's Being is an issue for it; the disclosive self-attunement that such moods exemplify confirms Heidegger's earlier claim that Dasein's capacity to encounter objects as ready-to-hand involves grasping them in relation to its own possibilities-for-Being. On the other hand, however, Dasein's Being is put at issue here by something in the world that is genuinely fearsome, that poses a threat to the person who fears. This reveals not only that the world Dasein inhabits can affect it in the most fundamental ways, that Dasein is open and vulnerable to the world, but also that things in the world are really capable of affecting Dasein. The threat posed by a rabid dog, the sort of threat to which Dasein's capacity to respond to things as fearful is attuned, is not illusory.

This argument against what might be called a projectivist account of moods is reminiscent of one developed by John McDowell.[3] In essence, the projectivist is struck by the fact that, when we characterize something as boring or fearful, we do so on the basis of a certain *response* to it, and she concludes that such attributions are simply projections of those responses. But, in so doing, she overlooks the fact that those responses are to things and situations *in the world*, and any adequate explanation of their essential nature must take account of that. So, for example, any adequate account of the fearfulness of certain objects must invoke certain subjective states, certain facts about human beings and their responses. It must also, however, invoke the object of fear – some feature of it that prompts our fear-response: in the case of a rabid dog, for example, the dangerous properties of its saliva. Now, of course, that saliva is dangerous only because it interacts in certain ways with human physiology, so invoking the human subject is again essential in spelling out what it is about the dog that makes it fearful; but that does not make its fearfulness any less real – as we would confirm if it bit us.

The point is that there are two senses in which something might be called subjective: it might mean 'illusory' (in contrast with veridical), or 'not comprehensible except by making reference to subjective states, properties or responses' (in contrast with phenomena whose explanation requires no such reference). Primary qualities like length are not subjective in either sense; hallucinations are subjective in both senses; and fearfulness (like secondary qualities and moral qualities, in McDowell's view) is subjective only in the second sense. In other words, whether something is really fearful is, in an important sense, an objective question – the fact that we can find some things fearful when they do not merit that response (e.g. house spiders) shows this; and in so far as our capacity to fear things permits us to discriminate the genuinely fearful from the non-fearful, then that affective response reveals something about the world.

Moreover, the relation of moods to those undergoing them – what we have been calling the subjective side of the question of moods – is not to be understood in an unduly subjective way. For Heidegger, since Dasein's Being is Being-with, its individual states not only affect but are affected by its relations to Others. This has two very important consequences. First, it implies that moods can be social: a given Dasein's membership of a group might, for example, lead to her being thrown into the mood that grips that group, finding herself immersed in its melancholy or hysteria. This point is reinforced by the fact that Dasein's everyday mode of selfhood is the they-self: 'Publicness, as the kind of Being that belongs to the "they", not only has in general its own way of having a mood, but needs moods and "makes" them for itself' (BT, 29: 178). A politician determining judicial policy on the back of a wave of moral panic is precisely responding to the public mood.

The socialness of moods also implies that an individual's social world fixes the range of moods into which she can be thrown. Of course, ontically speaking, an individual is capable of transcending or resisting the dominant social mood – her own mood need not merely reflect that of the public; but, even if it does not, the range of possible moods open to her is itself socially determined. This is because Dasein's moods arise out of Being-in-the-world, and that world is underpinned by a set of socially defined roles, categories and concepts; but it means that the underlying structure even of Dasein's seemingly most intimate and personal feelings and responses is socially conditioned.

This Heideggerian idea underpins Charles Taylor's notion of human beings as self-interpreting animals.[4] Taylor follows Heidegger's tripartite analysis of moods, arguing that an emotion such as shame is related in its essence to a certain sort of situation (a 'shameful' or 'humiliating' one), and to a particular self-protective response to it (e.g. hiding or covering up). Such feelings thus cannot even be identified independently of the type of situations that give rise to them, and so can be evaluated on any particular occasion in terms of their appropriateness to their context. But the significance of the term we employ to characterize the feeling and its appropriate context is partly determined by the wider field of terms for such emotions and situations of which it forms a part; each such term derives its meaning from the contrasts that exist between it and other terms in that semantic field. For example, describing a situation as 'fearful' will mean something different according to whether or not the available contrasts include such terms as 'terrifying', 'worrying', 'disconcerting', 'threatening', 'disgusting'. The wider the field, the finer the discriminations that can be made by the choice of one term as opposed to another, and the more specific the significance of each term. Thus, the significance of the situations in which an individual finds herself, and the import and nature of her emotions, is determined by the range and structure of the vocabulary available to her for their characterization. She cannot feel shame if she lacks a vocabulary in which the circle of situation, feeling and goal characteristic of shame is available; and the precise significance of that feeling will alter according to the semantic field in which that vocabulary is embedded.

It is not that the relationship between feeling and available vocabulary is a simple one. In particular, thinking or saying does not make it so: not any definition of our feelings can be forced upon us, and some that we gladly take up are inauthentic or deluded. But neither do vocabularies simply match or fail to match a pre-existing array of feelings in the individual; for we often experience how access to a more sophisticated vocabulary makes our emotional life more sophisticated. And the term 'vocabulary' here is misleading: it denotes not just an array of signs, but also the complex of concepts and practices within which alone those signs have meaning. When one claims that, for example, no one in early twenty-first-century Britain can experience the pride of a Samurai warrior because the relevant vocabulary is unavailable, 'vocabulary' refers not just to a set of Japanese terms but to their role in a complex web of customs, assumptions and institutions. And, because our affective life is conditioned by the culture in which we find ourself, our being immersed in a particular mood or feeling is revelatory of something about our world – is cognitively significant – in a further way. For then our feeling horrified, for example, not only registers the presence of something horrifying in our environment; it also shows that our world is one in which the specific complex of feeling, situation and response that constitutes horror has a place – a world in which horror has a place.

This is why Taylor and Heidegger claim that the relationship between a person's inner life and the vocabulary available to her is an intimate one. And, since that vocabulary is itself something the individual inherits from the society and culture within which she happens to find herself, the range of specific feelings or moods into which she may be thrown is itself something into which she is thrown. How things might conceivably matter to her, just as much as how they in fact matter to her at a given moment, is something determined by her society and culture rather than by her own psychic make-up or will-power. It is this double sense of thrownness that is invoked when Heidegger says: '*Existentially, a state-of-mind implies a disclosive submission to the world, out of which we can encounter something that matters to us*' (BT, 29: 177).

If states-of-mind reveal Dasein as thrown Being-in-the-world, understanding reveals it as carrying forward that momentum; it corresponds to the active side of Dasein's confrontation with its own existentiell possibilities. For, if Dasein's Being is an issue for it, then each moment of its existence it must actualize one of the possibilities which its situation makes available to it, or fail to do so and thereby fall into one of those possibilities (including, of course, the possibility of remaining in the state in which it finds itself). In other words, Dasein must project itself on to one or other existen-tiell possibility; and this projection is the core of what Heidegger means by 'understanding'. But any such projection both presupposes and constitutes a comprehending grasp of the world within which the projection must take place. It involves grasping the possibilities for practical action which that specific situation allows, and so grasping the world in relation to Dasein's own possibilities-for-Being. Just as with states-of-mind, then, understanding is a matter of comprehending the world as a context of assignments or references, a totality in which any given object relates to other objects and ultimately to a possibility of Dasein's Being:

> In the way in which its Being is projected both upon the 'for-the-sake-of-which' and upon significance (the world), there lies the disclosedness of Being in general. Understanding of Being has already been taken for granted in projecting upon possibilities . . . though not ontologically conceived.
>
> (BT, 31: 187)

It is easier to accept that projective understanding has a genuinely cognitive dimension than that moods possess an epistemology; but that makes it all the more important to understand the nature of the knowledge involved. As we saw when we analysed readiness-to-hand, this knowledge is essentially practical, a matter of know-how rather than knowing that: understanding is a matter of being competent to do certain things, to engage in certain practices. And this practical competence is essentially related to certain existentiell possibilities. How I relate to the objects around me is determined by the task for the sake of which I am acting (e.g. making a chair), but I perform that task for the sake of some more general existen-tiell possibility (e.g. being a conscientious carpenter) that serves to define who I am. In this way, the more general for-the-sake-of-which directs and constrains the more local. My self-understanding shapes the way in which I carry out – project myself upon – the more localized tasks with which I am confronted. More precisely, projecting myself in a particular way upon the latter just *is* to project myself in a particular way upon the former. But, then, living as a carpenter means continually projecting oneself in a certain way. One is at present a carpenter because one projected oneself on to that possibility in the past and, in the absence of such continued projection, the present substance of one's existence as a carpenter would dissolve. And that in turn implies that Dasein's true existential medium is not actuality but possibility:

> [A]ny Dasein has, as Dasein, already projected itself; and as long as it is, it is projecting. As long as it is, Dasein always has understood itself and always will understand itself in terms of possibilities. . . . As projecting, understanding is the kind of Being of Dasein in which it *is* its possibilities as possibilities.

> (BT, 31: 185)

Here, the question of authenticity re-emerges. For, in choosing to actualize one existentiell possibility rather than another, Dasein can either project itself upon a mode of existence through which its individuality can find proper expression (through which it can 'become what it is') or entirely fail to do so ('fail to find itself', perhaps by allowing the they-self to determine its choices, perhaps by [mis]understanding itself in terms of the categories appropriate to entities within its world – so that it loses its sense that finding itself is even a possibility). In short, projective understanding can be either authentic or inauthentic, although it is typically the latter; but projective inauthenticity is no less ontologically real than its authentic counterpart. Losing oneself or failing to find oneself are no less modes of Dasein's selfhood than finding oneself; if Dasein's Being is Being-in-the-world, then its understanding itself in terms of that world cannot amount to losing touch with itself ontologically.

The human capacity for projection is not, of course, entirely unanchored or free-floating. A particular Dasein cannot project itself upon any given existential possibility at any given time. First, the context might actually make it very difficult or even impossible to live in the way to which one has committed oneself: the conscientious carpenter may find herself working in a factory which entirely ignores the conceptions of good work by which she wishes to live. Second, someone who wishes to take on a certain social role may lack the necessary talents, or never be offered the necessary educational opportunities, or find herself in a state-of-mind in which a presented opportunity no longer possesses the attractions it once seemed to have. And, third, the range of existential possibilities upon which someone can project is determined by their social context. I could no more understand myself as a carpenter in a culture that lacked any conception of working with wood than I can understand myself as a Samurai warrior in early twenty-first-century Europe.

This shows that understanding always has only a relative autonomy; our projective capacities are as conditioned as our affective states. The freedom to actualize a given existential possibility is real but it is not absolute, since what counts as a real possibility is and must be shaped by the concrete situation and the cultural background (and their respective prevailing moods) within which the decision is taken, and these factors are largely beyond the control of the individual concerned. As Heidegger puts it:

> In every case Dasein, as essentially having a state-of-mind, has already got itself into definite possibilities. As the potentiality-for-Being which it is, it has let such possibilities pass by; it is constantly waiving the possibilities of its Being, or else it seizes upon them and makes mistakes. But this means that Dasein is Being-possible which has been delivered over to itself – *thrown possibility* through and through.
> (BT, 31: 183)

Dasein always faces definite possibilities because it is always situated (in the world). No situation reduces the available possibilities to one, but, unless a situation excluded many possibilities altogether, it would not be a situation (a particular position in existential space) at all. Just as thrownness is always projective (disclosing the world as a space of possibilities that matter to us in specific ways), so projection is always thrown (to be exercised in a field of possibilities whose structure it did not itself project). These are in fact two analytically separable faces of a single ontological structure; Dasein is thrown projection, and as such is subject to limits that must not be understood as limitations because one cannot conceive of any mode of human existence that lacked them.

If, however, we further explore the ontological underpinnings of understanding, we will see that it does not just essentially relate Dasein to the realm of possibility; it too has such a relation – our capacity for projective understanding itself possesses certain possibilities of self-development and self-realization. And, when they are actualized, those possibilities provide an important mode of access to the precise ontological structure of the capacity, and so to that of the being whose capacity it is.

Sometimes, the smooth course of our everyday activities is disrupted – when, for example, we are forced to stop in order to repair a broken tool, or to adapt an object for a given task, or even when a sudden access of curiosity leads us to contemplate an item in our work-world. In so doing, we engage in what Heidegger characterizes as 'interpretation', and the structures of our everyday comprehending engagement with these objects thereby become our explicit concern. Such interpretation is not something superimposed upon our practical comprehension, but is rather a development of it – the coming to fruition of a possibility that is inherent in projective understanding but which is not necessary for its usual, more circumspect functioning. In interpretation, we might say, the understanding appropriates itself understandingly, taking a practical interest in how it guides practical activity. And what then comes explicitly into sight is the following:

> All preparing, putting-to-rights, repairing, improving, rounding-out, are accomplished in the following way: we take apart in its 'in-order-to' that which is circumspectively ready-to-hand and we concern ourselves with it in accordance with what becomes visible through this process. That which has been circumspectively taken apart with regard to its 'in-order-to' and taken apart as such – that which is explicitly understood – has the structure of *something as something*.
> (BT, 32: 189)

This connection between seeing something *as* something and projective understanding is obvious in retrospect, for the types of category 'as which' we see things (as doors, hammers, pens) are of course specifications of the ways in which they can be woven into Dasein's practical activities. Seeing-as is simply the fundamental structure of the totality of reference- or assignment-relations that make up the world. But it also specifies how objects in the world make themselves intelligible to Dasein: it elucidates their fundamental significance or meaningfulness. In other words, Dasein's projective understanding and the intelligibility of ready-to-hand objects are related in just the way the concept of seeing-as is bound up with that of being-seen; they are two aspects of the same thing. The foundation or ground of Being-in-the-world is thus a unified framework or field of meaning with a very specific nature.

Once again, Heidegger is rejecting any interpretation of the world as essentially meaningless and of our relation to it as a matter of projecting subjective values or meanings upon it. To the Cartesian model of a present-at-hand subject juxtaposed with a present-at-hand object, he opposes his conception of Dasein as essentially worldly or environed, and of meaning as belonging to the articulated unity of Being-in-the-world:

> In interpreting, we do not, so to speak, throw a 'signification' over some naked thing which is present-at-hand, we do not stick a value on it; but when something within-the-world is encountered as such, the thing in question already has an involvement which is disclosed in our understanding of the world, and this involvement is something which gets laid out by the interpretation.
>
> (BT, 32: 190–1)

And what the interpretation lays out is the fact that it is always already grounded in a particular conceptualization of the object of our interests. We conceive of it in some particular way or other (our fore-conception), a way which is itself grounded in a broader perception of the particular domain within which we encounter it (our fore-sight), which is in turn ultimately embedded in a particular totality of involvements (our fore-having). The example of the broken tool illustrates the idea. When we stop to repair a hammer, our grasp of it as needing a particular modification emerges from our broader grasp of the particular work environment to which it must be restored, which is itself grounded in our basic capacity to engage practically with the world of objects. Similarly, my interpretation of this passage in *Being and Time* presupposes my interpretation of the book as a whole, and that interpretation is in turn guided by my particular interests in philosophy and my conception of what philosophy is, and so is ultimately dependent upon my assimilation of that particular facet of modern Western culture.

Whether or not this multiple embedding has *three* basic layers or aspects is unimportant. What matters is that there can be no interpretation (and, so, no understanding) that is free of preconceptions, and that this is not a limitation to be rued but an essential precondition of any comprehending relation to the world. The second part of this claim is what gives Heidegger's position its bite: for it opposes him not only to any interpreter who claims to have achieved, or even to be aiming at, a reading of a text that is entirely untainted by preconceptions, but also to any critic of an interpretation who takes the mere fact that it depends upon a preconception to demonstrate its prejudiced or distorted nature. If all interpretation necessarily involves preconceptions, the relevant task of such a critic is not simply to determine their presence in any particular case, but to evaluate their fruitfulness or legitimacy. On Heidegger's account, such evaluations will themselves be based on preconceptions, which must in turn be open to evaluation, and so on; but if this is taken to demonstrate the existence of a vicious circle, then understanding has been misunderstood from the ground up:

> What is decisive is not to get out of the circle but to come into it in the right way. This circle of understanding is not an orbit in which any random kind of knowledge may move; it is the essential *fore-structure* of Dasein itself. ... In the circle is hidden a positive possibility of the most primordial kind of knowing. To be sure, we genuinely take hold of this possibility only when, in our interpretation, we have understood that our first, last and constant task is never to allow our fore-having, fore-sight and fore-conception to be

presented to us by fancies and popular conceptions, but rather to make the scientific theme secure by working out these fore-structures in terms of the things themselves.

(BT, 32: 195)

No interpretation of an object could conceivably be free of precon-ceptions, because, without some preliminary orientation, however primitive, it would be impossible to grasp the object at all: we would have no sense of what it was we were attempting to interpret. But this does not mean that all interpretations are based on prejudice: for it is always possible to uncover whatever preconceptions we are using and subject them to critical evaluation. For example, with respect to this interpretation of Heidegger, we might ask how it is anchored in identifiable features of the text, whether a particular understanding of what philosophy is – an understanding which may perhaps lead us to reject Heidegger's work as philosophy – should not in fact be put in question by that work, and so on. The point is that we can and do distinguish between good and bad interpre-tations, and between better and worse preconceptions. We can only do so by allowing text, interpretation and preconception to ques-tion and be questioned by one another, but that essentially circular process can be virtuous as well as vicious. In short, there is a differ-ence between preconceptions and prejudices, and we can tell the difference.

This is not just a point about interpretations of texts – of literary criticism, Bible studies, history and the like. For Heidegger, it also applies to every sphere of human knowledge, the natural sciences and mathematics included: as aspects of Dasein's comprehending relation to the world, they must presuppose the fore-structure of understanding, which is simply more evident in the human sciences. Even mathematicians can approach their business only if they have some preliminary conception of what that business is – how it is to be conducted, what its standards of achievement are, which of its technical resources are legitimate and so on. Mathematicians may draw upon a very different, and less broad, totality of involvements than do students of history, but their efforts are no less based upon a prior comprehending grasp of the world: 'Mathematics is not more rigorous than historiology, but only narrower, because the existential foundations relevant for it lie within a narrower range' (BT, 32: 195). In short, in so far as interpretation lays bare the structures of understanding, it reveals something about *every* aspect of Dasein's existence in the world.

NOTES

1 See P. Strawson, *Individuals* (London: Routledge and Kegan Paul, 1959).
2 See L. Wittgenstein, *Philosophical Investigations* (Oxford: Blackwell, 1953), sections 185–243.
3 See J. McDowell, 'Values and Secondary Qualities', in T. Honderich (ed.) *Morality and Objectivity: Essays in Honour of J. L. Mackie* (London: Routledge, 1985).
4 See the works cited in the Introduction, note 4.

3

LANGUAGE, TRUTH
AND REALITY

(Being and Time, §§33–4, 43–4)

So far, Heidegger's account of the human way of being has isolated several of its defining limits or conditions – Dasein's worldliness; its communality; and its thrown projectiveness. It has also sketched in their interconnectedness – Dasein's world being intersubjectively structured and determinative of the available range of individual passions and projects. However, this picture of human conditionedness needs one further element, an element that derives from and determines the communal structures of Dasein's world – language. And Heidegger's analysis of language generates a distinctive account of the nature of truth and reality – one that overturns some of the pivotal assumptions of the post-Cartesian philosophical tradition. We will therefore break off from a purely linear treatment of Heidegger's text and devote this chapter to the two separate sequences of sections in which he examines these complex and tightly intertwined matters.

LANGUAGE: ASSERTIONS AND DISCOURSE (§§33–4)

The topic of language follows naturally on from Heidegger's treatment of understanding and interpretation, because the linguistic phenomenon of assertion is intimately connected with both. More precisely: just as interpretation is grounded in understanding, so assertion is grounded in interpretation; it is a species of that genus, but an extreme or specialized example of it.

Heidegger defines an assertion as 'a pointing-out which gives something a definite character and which communicates' (BT, 33: 199). Assertions therefore partake of the structures manifest in wordless interpretative activities such as repairing a tool. Considering how to modify a hammer so as to return it to use involves an interpretative fore-structure that brings to light the fore-structure of our understanding of it in use. Similarly, if we describe our difficulty – by saying 'The hammer is too heavy' – we pick out an object as having a certain character, thereby articulating a specific fore-conception of it which is recognizably related to the fore-structure of our wordless attempts to modify it (our focus upon a particular feature of the hammer), as well as the particular fore-sight and fore-having in which those efforts were embedded. Our assertion thus has a structure of the same type as that which grounded our original practical interaction with the object and was appropriated more explicitly in our subsequent interpretation of it. 'Like any interpretation whatever, assertion necessarily has a fore-having, a fore-sight and a fore-conception as its existential foundations' (BT, 33: 199).

By giving expression to our fore-conception of the object, we make it more broadly available; after all, assertions are usually made to communicate something to others. In this way, assertoric speech acts reflect the fact that Dasein's Being is Being-with. But, according to Heidegger, assertion also narrows down the focus of our concerns:

> In giving something a definite character, we must, in the first instance, take a step back when confronted with that which is already manifest – the hammer that is too heavy. In 'setting down the subject',

we dim entities down to focus on 'that hammer there', so that by
thus dimming them down, we may let that which is manifest be seen
in its own definite character as a character that can be determined.

(BT, 33: 197)

Making an assertion about an object restricts our openness to it in
just the way that interpretation restricts our pre-interpretative
understanding. When a tool needs repair, our grasp of an object
as ready-to-hand in an equipmental totality is narrowed down to
the object itself now understood as unready-to-hand. And, when we
encapsulate some information about what makes it unready-to-hand
for the benefit of others, we further restrict our concern to a specific
occurrent property of an object now understood as present-at-hand.
In short, such assertions are, if not theoretical, at least proto-
theoretical; they transform our relation to the object by severing it
from its place in a work-world of practical concern and situating it
solely as a particular thing about which a particular predication can
be made. As Heidegger puts it, 'our fore-sight is aimed at something present-at-hand in what is ready-to-hand' (BT, 33: 200); in a
single movement, what is ready-to-hand is covered up and what is
present-at-hand is discovered.

Thus, linguistic meaning (as manifest in assertion) is doubly
distanced from meaning per se – the field of significance that grounds
the human understanding of the world. Despite sharing the basic
structure of all understanding, an assertion's fore-conception of
entities as present-at-hand subjects of predication reductively transforms the interpretative fore-conception of entities as unready-to-
hand in some particular way, which itself is a restriction of our
pre-interpretative understanding of entities as part of a totality
of involvements. This gap is not, of course, unbridgeable. After
all, just as what interpretation grasps is nothing less than the fore-
structures of pre-interpretative understanding, so what assertions
articulate is what concerns us in our interpretations – that which
makes the given tool unready-to-hand. Assertions may tend to
disclose entities as present-at-hand, but it is a presence-at-hand
discovered 'in' their readiness-to-hand. Moreover, assertions modify
rather than annihilate the significance-structure of interpretation –

it dwindles or is simplified rather than being negated (cf. BT, 33: 200–1). Since making an assertion is a possible activity for Dasein, it is a mode of Being-in-the-world, and so grounded in the seeing-as structure that underpins the meaning of entities. Even with these qualifications, however, the meaning of assertions (narrow, reductive, levelling, decontextualizing) remains very different from the meaning that is articulated in the field of significance from which it ultimately derives. Accordingly, employing our understanding of assertions as a model or blueprint for human understanding of meaning per se could only result in error:

> It is not by giving something a definite character [in an assertion] that we first discover that which shows itself – the hammer – as such; but when we give it such a character, our seeing gets restricted to it.
>
> (BT, 33: 197)

Why, then, does Heidegger link language to the existential constitution of Dasein's disclosedness? After stressing that the foundational fore-structure of assertion covers up the totality of involvements and signification that underlies our understanding of the world, he immediately introduces the term 'Rede' (which means 'discourse', or, better, 'talk') as at once the existential-ontological foundation of language (including assertions) and the Articulation of intelligibility, claiming that 'the intelligibility of Being-in-the-world . . . *expresses itself as discourse*' (BT, 34: 204). Since assertion is reductive, 'discourse' must denote some *other* aspect of the existential-ontological foundations of assertoric (and, of course, non-assertoric) utterances, something genuinely disclosive of entities in their Being. But what might this be?

When we assert that a hammer is too heavy, this encourages a view of the hammer as an isolated present-at-hand entity because the subject–predicate structure of the assertion detaches it from its worldly environment, laying stress only on the question of whether or not it has a certain occurrent property. Even so, however, in making that assertion we use a linguistic term to categorize it as a particular kind of thing (namely, a hammer); to employ such a categorization, then, just *is* to see something as something – which is, of course, the foundational structure of significance or meaning, and so of practical understanding and interpretation. In short, the concepts and categories utilized in asserting something – what one might call the articulations of language – correspond to the articulations of the field of meaning. And this correspondence is not just a happy chance; rather, the inexplicit articulations of our understanding of the meaning of things, which are first appropriated explicitly in interpretations, find their most fitting fulfilment, their most explicit (and so in a sense most comprehending) appropriation, in recountings of the articulations that underlie language.

Heidegger's distinction between assertion and discourse might thus be understood as a distinction between a type of speech act and the conceptual framework upon which that speech act (along with every other speech act) must draw; and the latter can plausibly be thought of as the Articulation of the intelligibility of things. For, first, it is precisely a framework of meaning: it articulates the sense of the terms employed in specific speech acts to do certain things, and so functions as their enabling precondition. One could not assert that a hammer is heavy if the constituent terms of one's assertion had no meaning: only a grasp of that meaning allows one to pick out certain entities as hammers, and to determine whether they might correctly be described as heavy. Whether or not that assertion is true is determined by certain facts about the entity concerned. But any investigation of the world intended to make that determination must itself be guided by a grasp of what it is for something to count as a hammer and as heavy – and that does not itself derive from an investigation of the world (which would generate an infinite regress), but from a prior acquaintance with the conceptual framework of language. Nonetheless, since this framework articulates what it is for something to count as a specific type of entity, it specifies the essential nature of things: to know the criteria governing the use of the term 'hammer' just *is* to know what must be true of an entity if it is to count as a hammer, to appreciate the characteristics without which it would not be what it is. To grasp this framework is thus not just to grasp certain facts about our uses of words; it is also to grasp the essence of things. At this level, linguistic meaning and the meaning of entities are one and the same thing: the former discloses the latter, and thereby articulates the basis of Dasein's capacity to disclose entities in their Being.

None of this entails that language and discourse are identical. Rather, language – understood as a totality of words – is the worldly manifestation of discourse, the ready-to-hand (and sometimes present-at-hand) form of the Articulation of intelligibility. Discourse itself is not a worldly totality but an *existentiale* of Dasein, as much a facet of Dasein's disclosedness as are state-of-mind and understanding.

Consequently, the Being of discourse reflects these other facets of Dasein's Being. Since Dasein's Being is Being-with, language is essentially oriented towards others: it is a medium for communication, an essentially common inheritance from the culture or society in which a given Dasein finds itself thrown. This reflects one way in which discourse hangs together with state-of-mind; another lies in the way language is a medium within which Dasein expresses itself, giving utterance to its inner states or moods by the intonation, modulation and tempo of its talk. What reflects discourse's equiprimordiality with understanding is even more evident, in that language allows us to communicate about things in the world, to say something about something. In short, discourse, state-of-mind and understanding must be understood as three internally related aspects of Dasein's existential constitution – the three fundamental facets of its disclosedness, its Being-there.

REALITY AND TRUTH (§§43–4)

Since Dasein's capacity to disclose the Being of beings is the ontological underpinning of the human ability to grasp the true nature of reality, Heidegger's analysis of that capacity inevitably raises questions about reality and truth. More precisely, it raises the question of whether the concepts of reality and truth can be given an analysis adequate to their nature and yet consistent with the nature of Dasein. Heidegger's answer depends importantly upon the above account of the human relation to language.

In the modern Western philosophical tradition, 'reality' – understood as the realm of material objects deemed to exist 'outside' and independently of the human subject – appears as a problem: the problem is to demonstrate that reality is real, that there is such a world. But, for Heidegger, the real problem here is not that we have hitherto failed to demonstrate this, but that we persist in thinking that any such demonstration is needed: 'The "scandal of philosophy" is not that this proof has yet to be given, but that such *proofs are expected and attempted again and again*' (BT, 43: 249). For this expectation arises from a failure to comprehend properly the nature of Dasein's relation to its world, a failure that is itself based upon a misinterpretation of the Being of Dasein and the Being of 'the world'.

This misinterpretation is inevitably presupposed by any attempt even to state the problem of the external world. Those formulating it take for granted the existence of the human subject, and ask whether any of our beliefs about a world existing beyond our present moment of consciousness can be justified. But this presupposes that the human subject is such that the question of its own existence can coherently be bracketed off from the question of the existence of the world in which it dwells – and that conflicts with the fact that the Being of Dasein is Being-in-the-world. If, however, we think of persons not as essentially present-at-hand, immaterial substances but as inherently worldly, then it becomes impossible to state the problem of reality coherently; for the latter conception embodies precisely that transcendence of the 'sphere of consciousness' that is ineradicably problematic for the former. The same weakness emerges when the world whose existence is in question is conceptualized as an array of present-at-hand entities. If entities can only appear as such *within* a world, and if that world is founded upon the totality of assignment-relations that make up the worldliness of Dasein, then once again a proper ontological understanding of the world removes the logical distance between subject and world that is required to make their connectedness so much as questionable.

Heidegger's critique here does *not* take the form of answering the sceptic. On the contrary, if his analysis is correct, attempting to solve the Cartesian problem would be as fully misconceived as attempting to demonstrate its insolubility; the sceptic is no more deluded than the philosopher who aims to construct a refutation of scepticism. For a problem can be solved, and a question answered, only if problem and question can be stated coherently; so to treat a problem as requiring a solution, to regard a question as worthy of an answer, would amount to presupposing that they arise from an intelligible conception of their subject matter. If, then, we respond to the sceptic by asserting that the world really does exist, or that we can know of its existence with certainty, or that our certainty about its existence is based upon faith, we would be leaving unquestioned the terms of the Cartesian problematic and would thus reinforce rather than reject the misconceptions of subject and world that they presuppose.

We can see the point of this warning if we look a little more closely at the Cartesian conception of the relationship between subject and world. For, in formulating the 'problem of reality' as one of establishing whether we can know with certainty that the external world exists, and then claiming that this cannot be established, the sceptic presupposes that the 'relation' between subject and world is rightly characterized in cognitive terms, as one of knowing. As Heidegger points out, however, 'knowing is a *founded* mode of access to the Real' (BT, 43: 246), and is therefore doubly inapplicable as a model for the ontological relation between subject and world. First, because knowing is a possible mode of Dasein's Being, which is Being-in-the-world; knowing therefore must be understood in terms of, and so cannot found, Being-in-the-world. Second, because knowing is a relation in which Dasein can stand towards a given state of affairs, not towards the world as such; Dasein can know (or doubt) that a given chair is comfortable or that a particular lake is deep, but it cannot know that the world exists. As Wittgenstein might have put it, we are not of the *opinion* that there is a world: this is not a hypothesis based on evidence that might turn out to be strong, weak or non-existent.[1] Knowledge, doubt and faith are relations in which Dasein might stand towards specific phenomena in the world, but the world is not a possible object of knowledge – because it is not an object at all, not an entity or a set of entities. It is that within which entities appear, a field or horizon ontologically grounded in a totality of assignment-relations;

it is the condition for the possibility of any intra-worldly relation, and so is not analysable in terms of any such relation. What grounds the Cartesian conception of subject and world, and thereby opens the door to scepticism, is an interpretation of the world as a great big object or collection of objects, a totality of possible objects of knowledge, rather than as that wherein all possible objects of knowledge are encountered. And, for Heidegger, such an interpretation conflates the ontic and the ontological, assuming that a specific existentiell stance of the subject towards something encountered in the world might stand proxy for the *existentiale* that makes all such stances and encounters possible.

As we shall see in Chapter 4, this is not Heidegger's last word on the philosophical significance of scepticism. But, even if we restrict ourselves for the moment to this aspect of his strategy, it plainly presupposes the cogency of his analysis of Dasein's Being as Being-in-the-world; and since that classifies the worldhood of the world as an aspect of Dasein's ontological structure, it may seem to be open to the charge of subjectivizing reality, of quietly ceding its objectivity and independence while claiming to have preserved it from sceptical molestation. For, if the world is ontologically grounded in the Being of Dasein, must it not follow that when Dasein does not exist, neither does the world? And what reality is left to a world that is dependent for its own existence upon the continued existence of human creatures within it? If such a world is all that the Heideggerian analysis leaves us, is there any real difference between him and the sceptic?

This worry fails to take seriously the distinction between ontic and ontological levels of analysis in Heidegger's work. The significance of this omission is implicit in what he actually says about the matter:

> Of course, only as long as Dasein *is* (that is, only as long as an understanding of Being is ontically possible), 'is there' Being. When Dasein does not exist, 'independence' 'is' not either, nor 'is' the 'in-itself'. In such a case this sort of thing can be neither understood nor not understood. In such a case even entities within-the-world can be neither uncovered nor lie hidden. *In such a case* it cannot be said that entities are, nor can it be said that they are not. But *now*, as

> long as there is an understanding of Being and therefore an under-
> standing of presence-at-hand, it can indeed be said that *in this case*
> entities will still continue to be.
>
> (BT, 43: 245)

Note that Heidegger does not claim that 'entities exist only as long
as Dasein exists'; he claims that 'only as long as Dasein "is", "is
there" Being'. In other words, he invokes what he sometimes calls
the ontological difference; he distinguishes between entities and the
Being of entities, between material things and their nature and
actuality as things. But of what help is such a distinction?

Dasein encounters material things as phenomena that exist independently of its encounters with them. Part of what we mean when
we claim to see a table in the room is that we are seeing something
that was there before we entered the room and that will continue to
be there after we leave. Part of what we mean by 'the real world' is
a realm of objects that existed before the human species developed
and which is perfectly capable of surviving our extinction. In this
sense, to talk of objects just *is* to talk of real objects, objects which
exist independently of human thought and action; and we distinguish such things from such subjective phenomena as illusions, hallucinations and misleading appearances on the one hand, and from
moods, emotions and passions on the other – types of phenomena
which are dependent for their existence upon aspects of the human
constitution.

Accordingly, given what the term 'entity' means (what Heidegger
would describe as its what-being), it is simply incoherent to assert
that entities exist only as long as Dasein exists – for that amounts
to claiming that when Dasein is absent entities vanish, or that the
reality of a table in a room is dependent upon its being encountered
by a human creature. But, if Dasein were to vanish, then what would
vanish from the world would be the capacity to understand beings
in their Being, the capacity to uncover entities as existing and as
the entities they are. In those circumstances, it could not be asserted
either that entities exist or that they do not – for then there could
not be assertions about, or any other comprehending grasp of,
entities, any encounter with them in their Being.

We must distinguish between what can be said about entities-in-a-world-without-Dasein, and what can be said in-a-world-without-Dasein about entities-in-a-world-without-Dasein. Heidegger does *not* say: it cannot be said of entities existing in a world without Dasein that they exist (or that they do not exist). He says: in a world without Dasein, it cannot be said of entities that they exist (or that they do not exist). In so far as anything can be said about entities existing in such circumstances (i.e. in so far as there exists a being capable of assertion), then the only correct thing to say is that they will continue to exist as the entities they are; but, in those circumstances, it would not be possible to state anything, and so it could not be said either that entities continue to be or that they do not.

Heidegger underlines this distinction in the very way he formulates his position. For when he claims that 'only as long as Dasein *is*, "is there" Being', and that 'when Dasein does not exist, "independence" is not either', he deliberately encloses the crucial verbs in quotation marks. By simultaneously mentioning them and using them, he alerts us to the fact that the question of what it would be true to say about entities in a world without Dasein must not be conflated with the question of whether that truth could conceivably be uttered in such a world. And, by stressing the fact that truths are not just propositions that correspond to reality but the content of assertoric speech acts, he reminds us that an essential condition for the possibility of truth is the existence of Dasein.

In one sense of that claim, few would deny it. For it is trivially true that no truths could be enunciated in a world without creatures capable of enunciation; but the conditions for their enunciation are entirely independent of the conditions for their truthfulness – the latter simply being a matter of their fit with reality, something which the presence or absence of human creatures leaves entirely unaffected. But Heidegger means to claim something more. His point is that if truth is a matter of the correspondence between a judgement and reality, then the existence of Dasein is a condition for the possibility of truth – not because there can be no judgements without judgers, but because there can be no question of a judgement's corresponding (or failing to correspond) with reality without a prior articulation of that reality, and there can be no such articulation of reality without Dasein.

He discusses the case of someone who judges that 'the picture on the wall is askew'. After first stressing that the truth of this judgement is a matter of its corresponding to the picture itself and not to some mental representation of it, he then argues that what confirms its truth is our perceiving that the picture really is the way the judgement claims that it is:

> To say that an assertion 'is true' signifies that it uncovers the entity as it is in itself. Such an assertion asserts, points out, 'lets' the entity 'be seen' in its uncoveredness. The *Being-true* of the assertion must be understood as *Being-uncovering*. Thus truth has by no means the structure of an agreement between knowing and the object in the sense of a likening of one entity (the subject) to another (the Object).
>
> Being-true as Being-uncovering is in turn ontologically possible only on the basis of Being-in-the-world. This latter phenomenon . . . is the *foundation* for the primordial phenomenon of truth.
>
> (BT, 44: 261)

What is the basis for these claims?

Here we need to recall the distinction between assertion and discourse. An assertion is the utterance of a proposition, a statement that aims for truth: and whether it meets its aim is determined not by Dasein but by reality – by whether things are as it claims them to be. But in order for a proposition to be true or false – to fit or fail to fit its object – it must be meaningful. Before it can be determined whether it is true that the picture on the wall is askew, we must know what the terms 'picture', 'wall' and 'askew' mean. We must, in short, grasp the concepts of a picture, a wall and of spatial orientation from which that proposition is constructed. But to grasp those concepts, to understand the meaning of the relevant terms, one must be able to distinguish between correct and incorrect applications of them to reality – be able to grasp what (in reality) *counts* as a picture and what doesn't, and so on. So, these conceptual structures are not just articulations of language (what we earlier called 'discourse') but articulations of reality; in their absence, it simply would not be possible for a particular proposition to correspond or to fail to correspond to a particular piece of reality. The question of truth can only arise within the logical space created by a framework or field of meaning.

The opening up of this space of intelligibility is what Heidegger means by his talk of 'uncovering', which draws upon the Greek concept of truth as *a-letheia* (un-concealing). But if it is right to think of questions of truth as being settled within this space by assessing the correspondence between a proposition and its object, why does not the very same question arise with respect to the articulation of this logical space itself? What determines the validity of the framework of meaning if not its correspondence with the essential structures of the reality to which we apply it? Why then should Heidegger claim that uncoveredness is not a matter of correspondence?

Let's look again at the language side of the issue. The truth-value of a proposition may well be a matter of its correspondence with reality, but the significance of the conceptual categories in terms of which the proposition is articulated (i.e. the meanings of its constituent terms) are established by the norms or standards governing their use; and such norms do not stand in a relationship of correspondence (or of non-correspondence) with reality. Take the concept of water as an example, and assume that we define it as 'liquid with chemical composition H_2O'. That definition is not itself a claim about reality, something that might be true or false. It is the articulation of the following rule: if a liquid has the chemical composition H_2O, then it is water. It doesn't claim that any particular liquid does have that chemical composition, or that any such liquid is to be found anywhere in the universe. It simply licenses us to substitute one form of words ('water') for another form of words ('liquid with chemical composition H_2O'). It doesn't claim that the latter form of words is now, or is ever, applicable; it merely determines that, whenever that latter form of words is licitly applied, so is the former.

In other words, definitions are not descriptions, although they are an essential precondition for constructing descriptions since they confer meaning on the terms used in the description. In so far, then, as a conceptual framework is a specification of meanings (an articulation of intelligibility, in Heidegger's terminology), it simply is not a candidate for correspondence with reality. It does not embody a set of hypotheses or factual claims; rather, it determines what any given entity must have if it is to count as an instance of the relevant concept. It is not, therefore, possible for an examination of reality to show that our concepts fail to correspond to its essential nature; for any such examination would presuppose some framework or field of meaning, some set of categories in terms of which to describe what is discovered, and so could neither undermine nor justify that framework. The discovery that a given liquid does not have the chemical composition H_2O, or that there is no such liquid, would reveal not that our concept of water has misrepresented reality, but rather the local or global inapplicability of that concept. And, of course, if a conceptual framework is incapable of misrepresenting reality, it is also incapable of representing it accurately. Representation is not the business of concepts but of the empirical propositions constructed by deploying them; conceptual frameworks make correspondence between language and reality possible, but their relation to reality is not to be understood on the correspondence model.

Heidegger thinks of the human capacity to construct and apply concepts as manifesting our capacity to disclose entities because our conceptual framework embodies the fundamental categories in terms of which we encounter entities as entities of a particular sort, and indeed as entities (phenomena that continue to exist independently of our encountering them) at all. They determine the essential nature of phenomena in that they make manifest the necessary features of any given type of thing – those without which they would not count as an instance of that type at all; they articulate the seeing-as structure of meaning within which all encounters with entities must take place. But, if that structural aspect of language cannot be understood on the correspondence model, then it cannot be thought of as a discursive reflection of articulations in reality. Indeed, the very idea of reality as being already articulated in this way independently of discourse is incoherent. For, if the propositions that give expression to that structure do not state truths or falsehoods about reality, then the structure itself cannot be thought of as true or false to reality – which means that reality cannot coherently be thought of as inherently possessed of a structural essence to which these articulations of discourse might correspond, and which would exist in the absence of discursive creatures.

In other words, whereas the truth about reality must continue to hold even in the absence of Dasein, its essence cannot. The essential nature of reality is not simply one more fact about real things, one more aspect of the truth about the world that human beings come to know but which would continue to hold in their absence. Essence is not empirical, and so cannot persist independently of Dasein in the way that genuinely empirical matters do. The essentiality of a given feature of things – its status as necessary to the identity of the entity concerned – is not a function of the way things are in the world but of the way the conceptual framework is structured,[2] which is in turn dependent upon the field of meaning that underpins Dasein's understanding of entities in their Being. These articulations are thus ultimately ontologically grounded in Dasein's Being as Being-in-the-world.

Accordingly, a world without Dasein would not simply be a world without beings capable of making true judgements, but a world without the ultimate source of the categories in terms of which true and false judgements must be articulated, and so in which those articulations themselves are non-existent. It can and must be said (given our understanding of what it is to be an entity) that, in such circumstances, entities and the real world they make up would continue to exist. It could not be said, however, that Reality, Being or Truth would exist, for those terms denote reality in its essential nature, the articulation of the Being of things, the categorial conditions for the possibility of truth – and no sense can be attached to the idea that those articulations could exist in the absence of Dasein. It is this Truth with a capital 'T' to which Heidegger refers when he claims that ' "There is" truth only in so far as Dasein "is" and so long as Dasein "is" ' (BT, 44: 269), and that all truth is relative to Dasein's Being (*not* to Dasein):

> Does this relativity signify that all truth is subjective? If one Interprets 'subjective' as 'left to the subject's discretion', then it certainly does

not. For uncovering, in the sense which is most its own, takes asserting out of the province of 'subjective' discretion, and brings the uncovering Dasein face to face with the entities themselves. And only *because* 'truth', as uncovering, *is a kind of Being which belongs to Dasein*, can it be taken out of the province of *Dasein's* discretion. Even the 'universal validity' of truth is rooted solely in the fact that Dasein can uncover entities in themselves and free them. Only so can these entities themselves be binding for every possible assertion – that is, for every possible way of pointing them out.

(BT, 44: 270)

There can be no disclosure without Dasein; but what is disclosed are entities as they are in themselves, and so as the entities they always were before Dasein encountered them and the entities they will continue to be thereafter.

Nonetheless, if disclosure is the existential condition of the possibility of truth, and disclosedness is a mode or aspect of the Being of Dasein, then the most primordial understanding of truth is existential: Dasein *is* 'in the truth'. And, since Dasein is the kind of being whose Being is an issue for it, questions of authenticity and inauthenticity will apply to this mode of its Being as to all others. In other words, the being who alone can be said to be in the truth can also be in untruth; being capable of uncovering entities (including itself) as they are in themselves means that Dasein can fail to do so, can cover up the Being of beings. And which of those existential alternatives is that in which Dasein typically exists? Since we have had to overcome a strong philosophical tendency to treat the doubly derivative relation between present-at-hand propositions and states of affairs as the fundamental model for truth, in order to uncover a properly primordial understanding of it as rooted in disclosedness and existentiality, it seems that the inauthentic mode tends to prevail. But we need to examine the issue in more detail and in more generality. What is the everyday mode of Dasein's disclosedness, its Being-there?

NOTES

1   See L. Witttgenstein, *Philosophical Investigations* (Oxford: Basil Black-
    well, 1953), part 2, section iv, for a parallel remark about our relation
    to other people.
2   For a parallel view, see Wittgenstein's *Philosophical Investigations*,
    sections 371–3.

4

CONCLUSION TO DIVISION ONE: THE UNCANNINESS OF EVERYDAY LIFE

(Being and Time, §§34–42)

The question posed at the end of the previous chapter demands that we add a further element to the ontological web that constitutes Heidegger's account of the human way of being. It will show how average everyday social relations involve a particular kind of absorption in or preoccupation with the world, and so a particular kind of disclosure of it. But this addition permits Heidegger to conclude his preliminary investigation of human conditionedness by providing a single, overarching characterization of human existence that reveals the unity of its ontological underpinnings.

FALLING INTO THE WORLD (§§34–8)

Dasein, as Being-with, typically maintains itself in the Being of the they-self; so our question about Dasein's everyday mode of there-

Being amounts to asking how the they-self manifests itself from the perspective of disclosedness. Heidegger's answer focuses on three phenomena: idle talk, curiosity and ambiguity.

'Idle talk' is the form of intelligibility manifest in everyday linguistic communication – average intelligibility. All communication necessarily involves both an object (that which the conversation is about) and a claim about it. In idle talk, our concern for the claim eclipses our concern for its object. Rather than trying to achieve genuine access to the object as it is in itself, we concentrate upon what is claimed about it, taking it for granted that what is said is so, simply because it is said, and passing it on – disseminating the claim, allowing it to inflect our conversations about the object, and so on. We thereby lose touch with the ostensible object of the communication; our talk becomes groundless. And the ease with which we then seem to ourselves to understand whatever is talked about entails that we think of ourselves as understanding everything just when we are failing to do so. By suggesting such complete understanding, idle talk closes off its objects rather than disclosing them, and it also closes off the possibility of future investigations of them. An impersonal, uprooted understanding – the understanding of 'the they' – thus dominates Dasein's everyday relation to the world and Others.

An uprooted understanding of the world, detached from any particular task that might have focused Dasein upon objects in its immediate environment, tends to float away from what is ready-to-hand and towards the exotic, the alien and the distant. And if its focus is upon the novel, its primary concern tends to be with its novelty. It seeks new objects not in order to grasp them in their reality but to stimulate itself with their newness, so that novelty is sought with increasing velocity. In short, Dasein becomes curious: distracted by new possibilities, it lingers in any given environment for shorter and shorter periods; floating everywhere, it dwells nowhere. Being systematically detached from its environments, it cannot distinguish genuine comprehension from its counterfeit: superficial understanding is universally acclaimed as deep, and real understanding looks eccentric and marginalized. This ambiguity is not the conscious goal of any given individual; but, in a public world dominated by idle talk and curiosity, it permeates the understanding into which Dasein always already finds itself thrown, its inheritance from its fellows and its culture.

These three interconnected existential characteristics reveal a basic kind of Being that belongs to Dasein's everydayness – falling:

> This term does not express any negative evaluation, but is used to signify that Dasein is proximally and for the most part *alongside* the 'world' of its concern. This 'Absorption in . . .' has mostly the character of Being-lost in the publicness of the 'they'. Dasein has, in the first instance, fallen away from itself as an authentic potentiality for Being its Self, and has fallen into the 'world'.
>
> (BT, 38: 220)

In short, Dasein's average everyday disclosedness is inauthentic. Uprooted by its absorption in the 'they' from any genuine concern for its world and solicitude for its fellow human beings, it is also uprooted from any genuine self-understanding – any grasp of which possibilities are genuinely its own, as opposed to those which 'one' has.

This falling detachment from genuine self-understanding permeates Dasein's philosophical activities as well as those of its everyday life. Indeed, it constitutes Heidegger's central explanation for the fact that a being, to whom an understanding of its own Being naturally belongs, can nonetheless have a philosophical tradition which systematically represses any proper understanding of the human way of being. We saw earlier that philosophers tend to interpret the Being of Dasein in terms more appropriate to entities. We also saw that such misapplications of the category of presence-at-hand emerge naturally both from pre-theoretical absorption in our practical tasks (when objects lie temptingly ready-to-hand as paradigms of what it is for anything to exist), and from the peculiar circumstances of theoretical contemplation (in which both objects and human beings appear as entirely detached from their worlds). Dasein's inherent sociality and its tendency to lose itself in the 'they' suggested further that, once such misinterpretations were established in the philosophical culture, new generations of philosophers would tend unquestioningly to accept them as self-evident truths, as what everybody knows to be common sense. We can now see that philosophers who reject what is taken to be common sense in favour of ever more novel theoretical constructions, whose convolutions confer a thrill of the exotic or the intellectually advanced upon its proponents, are no less in thrall to the consensual hallucination of the they-world. Such philosophical inclinations are symptoms of a more general falling away from authentic self-concern and self-relation. Just as in other modes of human activity, philosophers become absorbed in the world of average everydayness because they have lost touch with themselves and with any awareness that they have a self with which they might lose touch.

But Heidegger does not just claim that falling is a general phenomenon – one to which any and every facet of human culture is always vulnerable. He also emphasizes that its ubiquity (and so the predominance of its effects in the philosophical tradition in particular) is not accidental. For, if falling is internally related to Dasein's absorption in the 'they', it must be just as much a part of Dasein's ontological structure as the they-self: falling is not a specific ontic state of Dasein, but 'a definite existential characteristic of Dasein itself' (BT, 38: 220). The ontological structures of Being-in-the-world do not make authenticity impossible; but neither do they leave the question of which specific ontic states Dasein might find itself in entirely open. If Dasein is always thrown into a world whose roles and categories are structured in inherently impersonal ways, in which idle talk, curiosity and ambiguity predominate, then absorption in the they-self will be its default position. It may then be able to find itself, but only by recovering itself from an original lostness. In this sense, authenticity always involves overcoming inauthenticity. 'In falling, Dasein *itself* as factical Being-in-the-world is something *from* which it has already fallen away' (BT 38: 220). The world into which Dasein finds itself thrown inherently tempts it to fall away from itself; and part of that fallen state, part of the ambiguity inherent in it, is a prevailing assumption that its fallenness is in reality fully authentic and genuine. The they-world thus tranquillizes Dasein; but this tranquillization finds expression in frenzied activity, a constant,

curiosity-driven search for the novel and the exotic, and a consequent alienation from the immediate environment and from
oneself – a self-alienation that sometimes takes the form of inces-
sant, curiosity-driven self-analysis. And this applies to Dasein's
philosophical activities as well: the various errors of self-
understanding to which the philosophical tradition is subject are
simply localized symptoms of this more general human state.

In short, then, Dasein's everyday state (within and without
philosophy) is one in which it finds itself thrown into inauthenticity: 'Dasein's facticity is such that *as long* as it is what it is,
Dasein remains in the throw, and is sucked into the turbulence of
the "they's" inauthenticity' (BT, 38: 223). It can achieve authenticity, but, when it does, it 'is only a modified way in which [falling]
everydayness is seized upon' (BT, 38: 224). Ontologically speaking,
authenticity is a modification of inauthenticity.

ANXIETY AND CARE (§§39–42)

One way of characterizing this average everydayness, Dasein's being
in untruth, would be as self-dispersal: Dasein is scattered amid the
constantly changing objects of its curiosity, caught up in the collection of selfless selves that make up the 'they', and fragmented by
its self-dissections. It is therefore curious that, up to this point,
Heidegger's analysis of Dasein's everydayness has suffered the same
fate. Although we are constantly reassured that Being-in-the-world
is a single, unified whole, we have so far been presented with what
seem like decontextualized fragments of that totality – the world,
Being-in, Being-with and Being-there – each itself subject to further
dissection. And just as an authentic mode of Dasein's existence
requires overcoming its self-dispersal, so a genuinely integrated
understanding of Dasein's Being requires gaining a perspective
on those fragments that demonstrates their overall unity. One
particular state-of-mind helps to solve both problems. As a mode
of existence, it forces inauthentic everyday Dasein to confront the
true structure of its existence; and, as an object of phenomenological analysis, it gives us access to a single unifying articulation of
Dasein's Being. That state-of-mind is anxiety or dread ('Angst').

Anxiety is often confused with fear. Both are responses to the world as unnerving, hostile or threatening, but, whereas fear is a response to something specific *in* the world (a gun, an animal, a gesture), anxiety is in this sense objectless. That in the face of which the anxious person is anxious is not any particular entity in the world. Indeed, the distinctive oppressiveness of anxiety lies precisely in its not being elicited by anything specific, so that we cannot respond to it in any specific way (e.g. by running away). For Heidegger, what oppresses us is not any specific totality of ready-to-hand objects but, rather, the *possibility* of such a totality: we are oppressed by the world as such – or, more precisely, by Being-in-the-world. Anxiety confronts Dasein with the knowledge that it is thrown into the world – always already delivered over to situations of choice and action which matter to it but which it did not itself fully choose or determine. It confronts Dasein with the determining and yet sheerly contingent fact of its own worldly existence.

But Being-in-the-world is not just that in the face of which the anxious person is anxious; it is also that *for which* she is anxious. In anxiety, Dasein is anxious about itself: not about some concrete existentiell possibility, but about the fact that its Being is Being-possible, that its existence necessarily involves projecting itself upon one or other possibility. In effect, then, anxiety plunges Dasein into an anxiety about itself in the face of itself. Since in this state particular objects and persons within the world fade away and the world as such occupies the foreground, then the specific structures of the they-world must also fade away. Thus, anxiety can rescue Dasein from its fallen state, its lostness in the 'they'; it throws Dasein doubly back upon itself as a being for whom its own Being is an issue, and so as a creature capable of individuality:

[I]n anxiety, there lies the possibility of a disclosure that is quite distinctive; for anxiety individualizes. This individualization brings Dasein back from its falling, and makes manifest to it that authenticity and inauthenticity are possibilities of its Being. These basic possibilities of Dasein (and Dasein is in each case mine) show themselves in anxiety as they are in themselves – undisguised by entities within-the-world, to which, proximally and for the most part, Dasein clings.

(BT, 40: 235)

By confronting Dasein with itself, anxiety forces it to recognize its own existence as essentially thrown projection, but its everyday mode of existence as fallen – completely absorbed in the 'they'. It emphasizes that Dasein is always in the midst of the objects and events of daily life, but that typically it buries itself in them – in flight from acknowledging that its existence (as Being-possible) is always more or other than its present actualizations, and so that it is never fully at home in any particular world.

Through this experience of uncanniness, anxiety lays bare the basis of Dasein's existence as thrown projection fallen into the world. Dasein's thrownness (exemplified in its openness to states-of-mind) shows it to *be* already in a world; its projectiveness (exemplified in its capacity for understanding) shows it to be at the same time ahead of itself, aiming to realize some existential possibility; and its fall-enness shows it to be preoccupied with the world. This overarching tripartite characterization reveals the essential unity of Dasein's Being to be what Heidegger calls *care* ('Sorge'):

> The formally existential totality of Dasein's ontological structural whole must therefore be grasped in the following structure: the Being of Dasein means ahead-of-itself-Being-already-in (-the-world) as Being-alongside (-entities-encountered-within-the-world). This Being fills in the signification of the term 'care'.

> (BT, 41: 237)

The proliferation of hyphens indicates that these provisionally sepa-rable elements of Dasein's Being are ultimately parts of a whole. And, by labelling that whole 'care', Heidegger evokes the fact that Dasein is always occupied with the entities it encounters in the world – concerned about ready-to-hand and present-at-hand entities, and solicitous of other human beings. The point is not that Dasein is always caring and concerned, or that failures of sympathy are impos-sible or to be discouraged; it is, rather, that, as Being-in-the-world, Dasein must *deal* with that world. The world and everything in it is something that cannot fail to matter to it.

Heidegger recounts an ancient creation myth, ostensibly to show that his interpretation of Dasein's nature is not unprecedented. In it, Cura shapes human beings from clay (donated by Earth) infused with spirit (donated by Jupiter); the three quarrel over its name, and Saturn determines that it shall be 'homo' (purportedly from 'humus', i.e. soil). This myth, however, is also a perspicuous representation of everything preceding it in the first division of *Being and Time* – an emblematic condensation of Heidegger's fundamental ontology of Dasein. For example, the temporal precedence of Cura's actions over those of Jupiter and Earth represents Dasein's Being as essentially unitary rather than compound, and as based in its concern for beings in their Being rather than in any one element of that putative compound. Nevertheless, the fact that Dasein is named after 'humus' suggests that the distinctively human way of being arises from its worldly embodiment rather than from any other-worldly capacity.

The myth also provides two other pointers that are important for our purposes. First, Cura's shaping of Dasein implies that Dasein is held fast or dominated by care throughout its existence. This signifies not only that care is the basis of its Being, but that this is something to which Dasein is subject – something into which it is thrown, and so something by which it is determined. After all, if Cura is Dasein's creator, then Dasein is the creature of care; and any creature is doubly conditioned – conditioned in that it is created rather than self-creating, and conditioned by the mode of its creation. Thus, in saying that Dasein is indelibly marked by its maker, the fable implies that care is the unifying origin of the various limits that characterize Dasein's distinctive mode of existence. So, by invoking this tale, Heidegger emblematizes the conditionedness of human existence – the human condition – as fundamentally a matter of being fated to a self and to a world of other selves and objects about which one cannot choose not to be concerned.

The fable's second lesson points forward rather than backward: as well as surveying what has gone before in *Being and Time*, it shows not only that more is to come, but also what that 'more' may be. For, of course, the character in the fable to whose authority even Cura must submit is Saturn; and Saturn is the god of Time. But if the creator of Dasein is herself the servant or creature of Saturn, then the most fundamental characterization of Dasein's Being must invoke not care but that which somehow conditions or determines care – time. In other words, Heidegger's invocation of this fable declares his conviction that uncovering care as the unifying onto-logical structure of human existence is itself only a provisional terminus for his existential analytic, and prepares the reader for the basic orientation of his investigations in Division Two – his sense that time, as that which conditions care, is itself the basic condition for the human way of being.

ANXIETY, SCEPTICISM AND NIHILISM

Before we move on to Division Two, however, I want to suggest that Heidegger's analysis of angst has a further moral – one which deepens our understanding of his relation to expressions of scepti-cism in philosophy. In Chapter 3, we saw that Heidegger considers it a scandal of philosophy that disproofs of scepticism about the external world are expected and attempted again and again; and this is because any proper conception of Dasein's worldliness makes the sceptic's questions inexpressible. Yet, as Heidegger's own formu-lation of the situation implicitly acknowledges, the scandal is apparently perennial – anti-sceptical expectations and attempts arise again and again, and a genuine understanding of the sceptical threat remains to be properly established in philosophy. Moreover, he has earlier recognized that, if the world is conceived of in Cartesian terms, sceptical doubts are not only articulable but also irrefutable; and such understandings of the world have pervasively informed the Western philosophical tradition, particularly in modernity. For Heidegger, then, scepticism is both evanescent and permanent: the sceptical impulse is certainly self-subverting (since its doubts annihilate a condition for the possibility of their own intelligibility) and yet also self-renewing (an apparently ineradicable human possi-bility which affects those possessed by it with a near-unshakeable faith in their own insight). How, then, should we understand this paradoxical state of affairs?

Since the sceptical stance is a particular human possibility, a way of understanding and grasping one's worldly existence, it must be analysable in terms of the existentialia Heidegger has identified in his analytic of Dasein; and that means in particular that it should be inflected by a particular mood. The true sceptic, as opposed to the straw figure of epistemology textbooks (and, as Heidegger says, 'perhaps such sceptics have been more frequent than one would innocently like to have true when one tries to bowl over "scepticism" by formal dialectics' [BT, 44: 272]), is someone beset by gnawing doubts: she is, in effect, in the grip of anxiety. Scepticism, one might say, just *is* how angst makes itself manifest in philosophy. But, as we have seen, Heidegger characterizes anxiety as a fundamentally revealing existentiell state, 'one of the the *most far-reaching and primordial* possibilities of [Dasein's] disclosure' (BT, 39: 226), in which Dasein reveals itself as a worldly being whose Being is an issue for it. So one should expect sceptical anxiety to embody exactly that kind of illumination. Does it?

For Heidegger, angst finds its clearest expression when someone gripped by it says that what makes her anxious 'is nothing and nowhere' (BT, 40: 231). This formulation highlights the fact that anxiety has no particular object – that neither that in the face of which one is anxious nor that about which one is anxious has a particular intra-worldly location. Anxiety is thus responsive to, and hence revelatory of, the world as such – that is, to the worldhood of the world, and thus to Dasein's own inherently worldly being. More specifically, it reveals Dasein as uncanny; it suggests that, at root, Dasein's way of Being-in-the-world is that of being not at home in the world. How might sceptical anxieties be thought to confirm or underwrite this paradoxical perception?

The 'external world' sceptic feels an abyss to open up between herself and the world, a sense of its insignificance or nothingness; she experiences a hollow at the heart of reality, and a sense of herself as not at home in the world. The 'other minds' sceptic feels an abyss to open up between herself and others, as if their thoughts and feelings were withdrawing unknowably behind their flesh and blood, as if she truly were confronted by hollowed out bodies, mere matter in motion; she experiences herself as alone in the world. In either mode, scepticism finds itself opposed to common sense, to the truths that average everyday human existence, with its absorption in phenomena and in the opinions of others, appears to confirm us in taking for granted; and, in this opposition, the sceptic at once falsifies and discloses the underlying realities of human existence. For, on Heidegger's account, we are essentially worldly, but we are also always more than any particular worldly situation in which we find ourselves; we are essentially Being-with, but we are also individuated. Hence, the intellectual (call it the traditional philosophical) expression of scepticism, in its argumentative denials of our worldliness and commonality, conceals the truth of Dasein's Being – as do familiar philosophical attempts to oppose those denials by argument; but the human anxiety of which philosophical scepticism is the intellectual expression, in its unwillingness to accept worldly absorption, reveals that truth.

Furthermore, the inarticulacy to which the sceptic's thwarted desire for connection with reality drives her makes manifest something vital about the discursive attunements upon which Dasein's capacity to grasp beings in their Being depends. For, if the sceptic can (however unknowingly) repudiate these articulations of meaning, then the common human attunement to the field of discourse must itself be contingent; the fact of scepticism shows that these articulations of meaning can exist only if Dasein continues to invest its interest or concern in them, and that Dasein can effect such withdrawals of interest in the guise of the most passionate investment of that interest. In other words, the self-subversiveness of scepticism shows that human responsiveness to the articulations of discourse, in which the issue of Dasein's own Being is most fundamentally at stake, is not something with which Dasein is automatically endowed – as if part of a pre-given essence that determines its existence. It is, rather, an inheritance for which Dasein must take (or fail to take) responsibility in and through its existence.

There is, however, a third aspect to the notion of Dasein's uncanniness that sceptical anxiety helps to bring out. For Heidegger previously showed that the worldhood of the world (to which anxiety as such is responsive) is a system of assignments of significance – a field of meaning; and he thereby suggested that the sense or meaning of our existence is ultimately to be understood as an aspect of Dasein's Being. And, if that is the case, then his analysis undercuts the possibility that the significance of our lives is anchored in a wholly external source or authority – whether that source is thought of as God, or as a range of Platonic Forms, or as a structure of values that is written into the independent reality of things in some other way. But how, then, can we regard the structures of significance that give orientation and meaning to our existence as having any genuinely objective authority, any real claim on us? Must they not be essentially anthropocentric constructions, designed to cover up the intrinsic meaninglessness of the world we inhabit – its inherent lack of sense? The anxious disclosure of the world as a domain in which we are ultimately not at home might then seem to be a wholly apt expression of this realization that the meaning of our lives lacks any external ground.

We might think of this aspect of Dasein's uncanniness as capturing the ontological root of what Nietzsche famously calls the problem of nihilism – that form of philosophical scepticism concerned with the reality or substance of value and meaning. But, once again, we will have to distinguish between the truth in such scepticism, and the falsity or distortions embodied in its intellectual expression. For, just as Heidegger argues in sections 43–4 that to acknowledge the internal relation between discourse and the Being of Dasein does not entail subjectivizing or relativizing our conceptions of truth and reality, so he seems committed to the claim that any authentic response to the problem of nihilism must find a way to acknowledge that life's meaning lacks any external grounding without denying its authoritative claims upon us. And the beginning of wisdom in this respect lies in seeing that, on his account of Dasein's Being, the very idea of a kind of meaningfulness that was wholly external in the relevant sense is empty.

Why? Because such an absolutely external structure of significance would have to be constituted in ways entirely independent of the ontological structure of Dasein's Being-in-the-world; but how then could it provide its inner articulation – how could it constitute the worldhood of the world, and thus orient and motivate Dasein's practical activities within it? On Heidegger's view, the thought that only a wholly external structure of meaning could make any authoritative claims on Dasein is the very reverse of the truth; it is, rather, that the only structures of meaning that could possibly make claims on Dasein are ones to which its worldly Being is inherently open, and by which it is articulated. In other words, the idea of objectivity that fuels nihilism does not specify a kind of authority that Dasein's fields of meaning could have, but unfortunately lack; it is the sheerest fantasy. But, if structures of significance could not conceivably be external in this sense. it cannot be right to think of the structures of significance in which we do and must exist as 'merely internal'. They are all the meaning there is, or could be, for creatures whose Being is that of Dasein; they are not limitations or constraints, but, rather, limits or conditions – essential determinations of any being whose Being is worldly, and hence finite.

The truth in nihilism is thus that Dasein's Being is essentially finite or conditioned; the truth is that Dasein is not unconditioned, not infinite or Godlike, and not entirely reducible to its determining conditions either. Dasein is not possessed of a wholly external ground, nor is it wholly self-grounding. Accordingly, in this respect as in the other two respects I specified earlier, to say that Dasein's worldliness is uncanny is to say that it must be understood in relation to nullity or negation, to what it is not and to that which is not – hence, in relation to nothing, or nothingness. This is the first (admittedly implicit and obscure) indication in Division One of a theme that will quickly come to full expression in the opening chapters of Division Two; and, in doing so, it radically alters our sense of what has been achieved in Division One as a whole. This, too, must inform our approach to the second half of *Being and Time*.

In all these ways, then, the sceptic truly suffers the reality of her existence as Being-in-the-world, even if she does not properly artic-ulate that reality, or make an issue of how her passionate anxiety might best be understood. That, however, is a vital part of the task of authentic phenomenology. As an activity engaged in by Dasein, phenomenological investigations of Being must be informed by some particular mood; and if the phenomenologist opens herself up to sceptical angst – if she not only subjects it to serious phenomeno-logical analysis, but also allows its unpredictable advent in her own existence to inform her sense of what matters in the distinctive field of her practical activity – then she will become receptive to the most far-reaching and primordial existentiell disclosure of the Being of Dasein. What could more properly facilitate her attempts to grasp Dasein's Being in as transparent a manner as possible – to make the existentiell possibility of investigating Dasein's Being truly her own?

But, of course, it is critical that the phenomenologist adopt a questioning attitude to her sceptical mood – and, in particular, that she not take scepticism's interpretation of its own significance for granted. She cannot, for example, accept the sceptic's over-anxious claim to know that the world is not knowable without acknowledging that the world cannot therefore be doubtable either. Authentically sceptical phenomenology will rather wrest the disclosures made possible by its own mood from that mood's self-concealments and dissemblings; it must overcome scepticism from within, by being sceptical about its self-understandings. It must, in short, dwell in this mode of Being-in-the-world without being at home in it. Only thus will it discover what is truthful about scepticism, and so what it is about scepticism to which philosophy must remain indebted.

5

THEOLOGY SECULARIZED: MORTALITY, GUILT AND CONSCIENCE

(Being and Time, §§45–60)

Heidegger's use of the ancient creation fable at the end of Division One ensures that his readers begin the second division of *Being and Time* knowing that its analysis of Dasein's underlying ontological structure will aim to connect the concept of care and that of time. It soon becomes clear that he wishes to forge that connection through a process of methodological self-reflection. He claims that his interpretation of the Being of Dasein hitherto – or, more precisely, its underlying fore-having or fore-sight – has been doubly restricted. First, by concentrating on Dasein's average everydayness, he has focused upon inauthentic modes of Dasein's Being to the detriment of its capacity for existentiell authenticity. And, second, by concentrating on the existential structure of specific moods and states of mind, he has downplayed the general structure of Dasein's life understood as a whole or a unity. Division Two makes good these omissions, and in a way which contributes to his overarching attempt to demonstrate the fundamentality of time to Dasein's Being. In effect, the tripartite thematic concern of Division Two is: authenticity, totality and temporality. This chapter follows Heidegger's initial development of the first two themes; the two following chapters examine his treatment of the third.

Given Heidegger's emphasis on the circular hermeneutic structure of understanding, it is natural to envisage Division Two as deepening our understanding of the claims made in Division One by drawing out their implications. The relevant image of their relation would be that of two turns around a spiral: each turn returns us to our starting point, but at a deeper level of ontological understanding, and each return opens the possibility of a new turn at a deeper level. Thus, Division One begins from a provisional conception of Dasein as the being who questions, and, by unfolding the articulated unity of the worldly existential structure implicit in that conception, it returns us to a deepened understanding of Dasein in terms of care; this is the first turn around the spiral. Division Two begins from that deepened conception of Dasein as care, and unfolds the articulated unity of temporality implicit in it, thus revealing that the care-structure presupposes an internal relation between the Being of Dasein and time; this is the second turn. The image of a spiral further incorporates Heidegger's rejection of the idea of absolute starting points and termini in human inquiry; for it implies that each new turn of ontological discovery presupposes its predecessors (and ultimately an initial leap into the circling process), and that the results of each turn will engender another turn.

Such an image of the book's progress is not exactly wrong; but it becomes clear by the end of the first two chapters of Division Two that it does not capture the full complexity of its internal structure. For the results of Heidegger's study of mortality, guilt and conscience do not simply deepen our understanding of the claims advanced in Division One and summarized in the characterization of Dasein's Being as care; by providing an uncanny background or horizon against which to re-articulate them, they also destabilize and even in a sense subvert them. It will be an important part of this chapter's business to try to understand the deep, but creative and even revelatory tension that this creates between the two Divisions of *Being and Time*.

DEATH AND MORTALITY (§§46–53)

Any philosophical attempt to grasp Dasein's existence as a totality or whole faces the problem that, in so far as Dasein exists, it is oriented towards the next moment of its existence and so is incomplete; but, once its existence has been brought to an end, once its life as a whole is over and so available for examination, Dasein itself is no longer there to prosecute that examination. In more existential terminology: Dasein always already projects upon possibilities, and so is oriented towards the not-yet-actual; so that structural incompletion is overcome only when Dasein becomes no-longer-Being-there. Thus, the idea of Dasein grasping its existence as a totality seems to be a contradiction in terms: for Dasein to be a whole is for Dasein to be no longer, and so to be no longer capable of relating to itself as a whole.

The problem is death. Death brings human existence to an end, and so completes it, but no one can experience her own death. As Wittgenstein put it, unlike dying, one's death is not an event in one's life – not even the last one.[1] It seems, therefore, that no Dasein can grasp its own existence as a whole. But this is not just a stumbling block for every human individual trying to make sense of her existence; it is a profound challenge to Heidegger's sense of what he has achieved in Division One, and of what he can achieve with his phenomenological method. For, remember, his concluding characterization of Dasein's Being as care in Division One was meant to allow us to grasp Dasein's Being as a whole, and thus provide a stable, even if provisional, resting-place for his existential analytic. But one aspect of the care-structure is Being-ahead-of-itself; and it is precisely this articulation – that is, Dasein's orientation towards the not-yet-actual – that hides within it the problematic of death, and hence conceals an essential incompleteness in the analysis. And the prospects of filling that analytical gap do not look at all promising, if one further recalls that Heidegger's phenomenological method relies upon Dasein's capacity to allow phenomena to disclose themselves as they are in themselves in its encounters with them. But we have just seen that no Dasein ever encounters its own death; so how, even in principle, could there be a genuinely phenomenological understanding of death, and so a genuinely complete existential analytic of Dasein?

Dasein can, of course, relate to the death of others, whether as dying or as dead. But this does not mean that we can grasp another's life as a totality, and thereby gain a proper understanding of the Being of Dasein in its wholeness. We can experience the transition from another Dasein's Being (-as-dying) to their no-longer-Being; we relate to their corpse as more than just a body – it is, rather, a body from which life has departed; and, as we can continue to relate to the dead person as dead – through funerals, rites of commemoration and the cult of graves – our lives after their death can involve modes of Being-with them (as dead, or no longer with us). But these are aspects of the significance of this person's dying and death to those of us still living; they are modes of *our* continued existence, not of theirs. To grasp the life of the dead person as a whole, we must grasp the ontological meaning of her dying and death *to her*; it is the totality or wholeness of *her* life that is at issue. Our access to the loss and suffering that this person's dying signifies for others brings us no closer to the loss-of-Being that she suffers, and so no closer to what it is for an individual Dasein's existence to attain wholeness or completion.

Nevertheless, this false trail carries an implication that will turn out to be crucial for our purposes, namely that no one can represent another with respect to her dying and death, that death is in every case ineliminably mine, unavoidably that of one particular individual. But before pursuing this, we must gain a more detailed understanding of the phenomenon of death and its role in the life of Dasein – uncover its existential significance. Death is the end of a person's life – but what sort of 'end'? Presumably, that in which Dasein's distinctive lack of totality finds its completion – but what sort of totality is that?

Death for Dasein is not a limit in the way that a frame is the limit of a picture or a kerbstone the limit of a road. The picture ends at the frame, but it is not annihilated by it in the way that death annihilates Dasein; the kerbstone marks the end of the road and the beginning of a new environment into which one can step from the road, whereas the death of the body is not another mode of its life. Such disanalogies demonstrate the futility of modelling any aspect of Dasein's existence on present-at-hand things; and ready-to-hand things are equally inappropriate. We might, for example, think of a human life as the accumulation of elements (moments, events, experiences) into a whole – as a sum of money is an accumulation of the coins and notes that make it up. Death then appears as the final element, the piece that completes the jigsaw. But, of course, when death comes to Dasein, Dasein is no longer there; life is no almost-complete edifice to which death can provide the coping stone.

The life of vegetable matter, of plants or fruit, might prove a better analogy: death would then signify the natural culmination of Dasein's existence in just the way that the mature state of a plant or the ripened state of a fruit completes its life cycle. But maturity is the fulfilment of the growing plant, just as ripeness is the end towards which the unripe fruit tends; whereas death is not the fulfilment of Dasein – Dasein may, and often does, die unfulfilled, with many of its distinctive possibilities unexplored, its *telos* unattained. The same is true of non-human animals: dogs and cats live and die, and they can often die without having actualized many of the possibilities that their nature leaves open to them. But Heidegger distinguishes sharply between the death of animals (which he calls their 'perishing') and that of Dasein. He acknowledges that Dasein is vulnerable to death in just the way that any living creature is so vulnerable, so that its biological or organic end (what Heidegger calls Dasein's 'demise' – cf. BT, 49: 291) is open to medical study. Even its demise, however, is not identical with the perishing of non-human animals, because Dasein's biological or organic identity is necessarily inflected by its distinctively existential mode of Being – in other words, by the fact that its life can be imbued with a knowledge of its own inevitable end, that it can relate to death as such. Dogs and cats must die, but that fact is only coded into their lives at the level of their species-identity. They strive to avoid death by obtaining nourishment and avoiding predators, and they contribute to the survival of their species by reproducing themselves. But these are not decisions that they take as individual creatures, but rather patterns of behaviour that they inherit and enact with as little consideration or awareness, as little scope for individual choice, as they have with respect to their bodily form.

In short, an animal's relation to death is as different from Dasein's relation to death as animal existence is different from human existence. Dasein has a life to lead, it exists – it must make decisions about which existentiell possibilities will be actualized and which will not. Death's true significance as the end of Dasein, as its completion or totalization, thus depends upon the significance of Dasein's existence as thrown projection, as a being whose Being is care. Hence, to understand death, we must attempt to undertand it existentially – that is, as one possibility of Dasein's Being. Since no Dasein can directly apprehend its own death, we must shift our analytical focus from death understood as an actuality to death understood as a possibility; only then can we intelligibly talk of death as something towards which any existing Dasein can stand in any kind of substantial, comprehending relationship. In other words, we must reconceive our relation to our death not as something that is realized when we die, but, rather, as something that we realize (or fail to) in our life.

What, then, is the distinctive character of this possibility of our Being, as opposed to any other (such as eating a meal, or playing football, or reading philosophy)? Heidegger gives us the following succinct summary:

> Death is the possibility of the absolute impossibility of Dasein. Thus death reveals itself as that *possibility which is one's ownmost, which is non-relational, and which is not to be outstripped*. As such, death is something *distinctively* impending.

> (BT, 50: 294)

Death impends, it stands before us as something that is not yet; but, unlike any other possibility of Dasein's Being, it can only stand before us. A storm or a friend's arrival can impend; but they can also arrive, be made actual. By contrast we cannot relate to our death as anything other than an impending possibility – for, when that possibility is actualized, we are necessarily no-longer-Dasein; death makes any Dasein's existence absolutely impossible. Hence, we can comport ourselves towards death only as a possibility; and, further, it stands before us as a possibility throughout our existence. A storm or a friend's arrival does not impend at every moment of our existence; but there is no moment at which our death is not possible – no moment of our existence that might not be our last. Hence, death – unlike any other possibility of Dasein's Being – is always and only a possibility; our fatedness to this purely impending threat makes concrete the articulated unity of our existence as thrown projection, our being always already delivered over to being ahead of ourselves.

Since what impends is Dasein's utter non-existence, and since Dasein must take over that possibility in every moment of its existence, Heidegger claims that, in relation to death, Dasein stands before its ownmost potentiality-for-Being – that possibility in which what is at issue is nothing less than Dasein's Being-in-the-world. Since Dasein is certain to die at some point, he further claims that death is a possibility that is not to be outstripped. And to complete his characterization, Heidegger (recalling his earlier claim that no one can take another's death away from her) also claims that, in Dasein's comportment towards its death, 'all its relations to any other Dasein have been undone' (BT, 50: 294) – in other words, that death is a non-relational possibility.

Of course, the non-relationality of death is hardly unique to it among our existential possibilities; if no one else can die my death, it is also true that no one else can sneeze my sneezes. However, sneezing fails to exemplify the other two elements in Heidegger's tripartite existential characterization of death (our very existence as Being-in-the-world is not at issue when we catch a cold, and at the very least it makes sense to imagine a human being who never sneezed). But, in another sense, it is precisely Heidegger's point that the non-relational nature of death highlights an aspect of Dasein's comportment to any and all of its existential possibilities; for, in making concrete Dasein's Being-ahead-of-itself, the fact that no one can die our death for us merely recalls us to the fact that our life is ours alone to live.

But, before examining this implication of Heidegger's analysis more closely, it is important to see that we have so far passed over a critical complication in Heidegger's approach to death. It may seem that, by treating death from an existential point of view – that is, as a possibility of Dasein's Being to which it must relate from within its existence – Heidegger has overcome death's obdurate resistance to any phenomenological grasp of its being. But such a conclusion would involve overlooking one remarkable feature of death understood as an existential possibility – the fact that it is not really an existential possibility at all. For any genuine existential possibility is one that might be made actual by the Dasein whose possibility it is: I might eat the meal I'm cooking, or play the game for which I'm training. But our own death cannot be realized in our existence; if our death becomes actual, we are no longer there to experience it. In other words, death is not just the possibility of our own nonexistence, of our own absolute impossibility; it is an impossible possibility – or, more frankly, an existential impossibility. But, if it amounts to a contradiction in terms to think of death as an existential possibility, of however distinctive a kind, then it would seem that Heidegger must be wrong to think that he can gain phenomenological access to death even by analysing it in existential terms.

This is where the real elegance of Heidegger's strategy for overcoming death's resistance to human understanding becomes clear. For, if death cannot coherently be regarded as even a very unusual kind of existential possibility (since an impossibility is not one genus of the species 'possibility', any more than nonsense is a kind of sense), then we cannot understand our relation to our own death on the model of our relation to any genuine possibility of our Being – as if our death stood on the same level (the ontic or existentiell level) as any other possibility upon which we might project ourselves. Heidegger's point in calling our relation to our own end our 'Being-towards-death' is precisely to present it as an ontological (that is, existential) structure, rather than as one existentiell state (even a pervasive or common one) of the kind that that structure makes possible. In short, we cannot fully grasp Heidegger's account of death except against the horizon of his account of the ontological difference – the division between ontic and ontological matters.

Why, then, call death an existential possibility at all? Doesn't this choice of terminology actually encourage forms of misunderstanding

that Heidegger must then attempt to avert – by, for example, emphasizing that an appropriately authentic relation to one's death is
not a matter of actualizing that possibility (say, by suicide), or of
expecting it to be actualized at every next moment, or of meditating
upon it in those terms? There is, however, a compensating and
fundamental advantage in Heidegger's view. For his terminology
underlines his key insight – namely, that, although we can't coherently regard death as an existentiell possibility, neither can we
understand our relation to our own end apart from our relation
to our existentiell possibilities, and thereby to our Being-ahead-of-
ourselves. More specifically, Heidegger's suggestion is that we
should think of our relation to death as manifest in the relation
we establish and maintain (or fail to maintain) to every genuine
possibility of our Being, and hence to our Being as such.

Precisely because death can be characterized as Dasein's ownmost,
non-relational and not-to-be-outstripped possibility, and, hence, as
an omnipresent, ineluctable but non-actualizable possibility of its
Being, which means that it is an ungraspable but undeniable aspect
of every moment of its existence, it follows that Dasein can only
relate to it in and through its relation to what *is* graspable in its
existence – namely, those genuine existentiell possibilities that
constitute it from moment to moment. Death thus remains beyond
any direct existential (and, hence, phenomenological) grasp; but it
is shown to be graspable essentially indirectly, as an omnipresent
condition of every moment of Dasein's directly graspable existence.
It is not a specific feature of the existential terrain, but, rather, a
light or shadow emanating evenly and implacably from every such
feature; it is the ground against which those features configure themselves, a self-concealing condition for Dasein's capacity to disclose
its own existence to itself as it really is.

In other words, just as Heidegger earlier reminded us that death
is a phenomenon of life, so he now tells us that death shows up
only in and through life, in and through that which it threatens
to render impossible – as the possible impossibility of that life.
Phenomenologically speaking, then, life is death's representative,
the proxy through which death's resistance to Dasein's grasp is
at once acknowledged and overcome, or, rather, overcome in and

through its acknowledgement. Death can be made manifest in our existential analytic only through a thorough recounting of that analysis in the light of the possible impossibility of that which it analyses. Or, to put matters the other way around: Being-towards-death is essentially a matter of Being-towards-life; it is a matter of relating (or failing to relate) to one's life as utterly, primordially mortal.

What might this amount to? Systematically transposing Heidegger's distinguishing predicates for death on to life, we might say the following. For Dasein to confront life as its ownmost possibility is for it to acknowledge that there is no moment of its existence in which its Being as such is not at issue. This discloses that Dasein's existence matters to it, and that what matters about it is not just the specific moments that make it up, but the totality of those moments – its life as a whole. Dasein thereby comes to see that its life is something for which it is responsible, that it is its own to live (or to disown) – that its existence makes a claim on it that is essentially non-relational, not something to be sloughed off on to Others. And to think of one's life as fated to be stripped out, rendered hollow or void, by death is to acknowledge the utter non-necessity of its continuation, and, hence, its sheer, thoroughgoing contingency. The hardest lesson of our mortality is its demand that we recognize the complete superfluity of our existence. Our birth was not necessary; the course of our life could have been otherwise; its continuation from moment to moment is no more than a fact; and it will come to an end at some point. To acknowledge this about our lives is simply to acknowledge our finitude – the fact that our existence has conditions or limits, that it is neither self-originating nor self-grounding nor self-sufficient, that it is contingent from top to bottom. But no representation of ourselves is harder to achieve or enact than this one; nothing is more challenging than to live in such a way that one does not treat what is in reality merely possible or actual or conditionally necessary as if it were absolutely necessary – a matter of fate or destiny beyond any question or alteration. Authentic Being-towards-death is thus a matter of stripping out false necessities, of becoming properly attuned to the real modalities of human existence.

This last perception is what most clearly connects Heidegger's project of representing Dasein to itself as a whole, and his desire to include the possibility of Dasein's authenticity in his general portrait of human everydayness. For an authentic grasp of Dasein's existence as mortal will inflect its attitude to the choices it must make (to its Being-ahead-of-itself) in four interrelated ways. A mortal being is one whose existence is contingent (it might not have existed at all, and its present modes of life are no more than the result of past choices), whose non-existence is an omnipresent possibility (so that each of its choices might be its last), a being with a life to lead (its individual choices contributing to, and so contextualized by, the life of which they are a part), and one whose life is its own to lead (so that its choices should be its own rather than those of determinate or indeterminate Others). In short, an authentic confrontation with death reveals Dasein as related to its own Being in such a way as to hold open the possibility, and impose the responsibility, of living a life that is genuinely individual and genuinely whole – a life of integrity, an authentic life.

But, of course, Heidegger does not think that Dasein typically does relate authentically to its own end, and hence to its own life. On the contrary: we typically flee in the face of death. We regard death as something that happens primarily to others, whom we think of as simply more cases or instances of death, as if they were mere tokens of an essentially impersonal type. We encourage the dying by asserting that it will never happen; and, on those occasions when it does, we often enough see it as a social inconvenience or shocking lack of tact on the deceased person's part – a threat to our tranquillized avoidance of death. Although we may never actually deny that it will happen to us, we are happy to contemplate courses of action that might promise to hold it off (whether temporarily, as with fitness schemes, or indefinitely, as with cryogenics); and we tend to regard it as a distant eventuality, as something that will happen but not yet, and hence as an impending event rather than as the omnipresent impending possibility of our own non-existence, that impossible but ineluctable possibility without which our existence would lack its distinctively finite significance.

This kind of tranquillizing alienation bears the characteristic marks of Dasein's average everyday existence in 'das man'; and it suggests that lostness in 'das man' is best understood as entanglement in a misplaced sense of the necessities of finite life. For it is part of this everyday mode of Dasein's Being that we regard the array of existential possibilities presently open to us, and the specific choices we make between them, as wholly fixed by forces greater than, or external to, ourselves. We do what we do because that is what one does, what is done, what 'das man' does; we displace our freedom outside ourselves, existing in self-imposed servitude to 'das man', unwilling not only to alter that fact but to acknowledge that it is a fact (but no more than that, an actuality and not a necessity). The reality is that we alone are responsible for allowing ourselves to be lost in the range of possibilities that our circumstances have thrust upon us, and we alone are capable of, and responsible for, altering that state of affairs.

This is why Heidegger characterizes authentic Being-towards-death as a mode of anxiously resolute anticipation. It is essentially anticipatory because death (the impossible possibility) can *only* be anticipated; and it is essentially anxious because to live in the light of a proper awareness of one's mortality is to make one's choices in the light of an extreme and constant threat to oneself that emerges unwanted and unbidden from one's own Being: it is to choose in the face of the nothing – the possible impossibility – of one's own existence. And, for Dasein to be oppressed by its own existence, by Being-in-the-world as such, just *is* – as we saw earlier – for Dasein to be anxious. And Heidegger's portrait of death as an ungraspable possibility reinforces this connection, by underlining the fit between death and the essential objectlessness of angst. For no object-directed state of mind could correspond to an existential phenomenon that utterly repels any objective actualization within Dasein's worldly existence; putting matters the other way around, to apprehend our worldliness as essentially uncanny, as a matter of not-at-homeness, just *is* to apprehend the mortality of our existence.

Here – in this conjunction of Dasein's non-necessity and its not-at-homeness – we can see the first appearance in Division Two of a theme which binds Heidegger's analysis of death together with

his analyses of guilt, conscience and temporality: the internal relation between Dasein and nothingness, nullity or negation. Our grasp
of its full significance must thus wait upon a proper account of the
remainder of Division Two. But, even at this early stage, we can
see that it suggests a rather more complex relationship between
Division Two and Division One than could be captured by the image
of two successive turns around a hermeneutic spiral. For that
image tends to suggest that Division Two simply deepens our grasp
of what is established in Division One – as if the issues broached
in Division Two simply take the articulated unity of the care-
structure entirely for granted, and concentrate on unfolding its
temporal implications. But, if death is essentially implicit in one
aspect of the care-structure (as well as in the mood that reveals that
structure), and if it lies essentially beyond direct phenomenological
representation, then it follows that to acknowledge death philosophically is to put in question our sense that the care-structure
gives us even a provisional grasp of Dasein as a whole, as well as
our sense that any such grasp is possible even in principle.

More precisely, in so far as Heidegger succeeds in attaining a
properly phenomenological grasp of death only by conceding the
impossibility of ever doing so, he implies that we cannot understand Dasein's Being without understanding that it is internally
related to that which lies beyond phenomenological representation.
He thereby invokes a new horizon or broader context for the whole
of his existential analytic of Dasein as presented in Division One –
the requirement to relate every element of it to that which is neither
a phenomenon nor of the logos, to that which (phenomenologically
speaking) cannot appear as such or be the object of a possible discursive act. For nothingness is not a representable something, and not
an unrepresentable something either; hence it can be represented
only as beyond representation, as the beyond of the horizon of the
representable – its self-concealing and self-disrupting condition.

Since this horizon is that of 'the nothing', then to invoke it as a
broader context for the analysis of Division One is in one sense to
add nothing whatever to that analysis – for it provides no specific
analytical ingredient in addition to those laid out in Heidegger's
initial characterization of the care-structure, and so nothing in

Division Two implies that this characterization is essentially incomplete. In another sense, however, introducing this relation to 'the nothing' as internal to Dasein's Being means introducing the thought that every element in the articulation of the care-structure is related to 'the nothing', and so must be reconsidered in its uncanny light. In that sense, by introducing this unthematizable theme of nothingness, Heidegger alters nothing, and everything, in his existential analytic.

One might say: if 'the nothing' really is the self-concealing and self-disrupting condition of Dasein's comprehending and questioning relation to Being, then phenomenological philosophy can only acknowledge it as such (that is, allow it to appear as it is) by allowing 'the nothing' first to conceal itself and then to disrupt its concealment *in the phenomenological analysis itself* – that is, to appear within the analysis as that upon which the analysis as a whole is shipwrecked. Only in this way could an existential analytic of Dasein achieve the kind of completeness that its condition allows and its object discloses – by presenting itself as essentially incomplete, beyond completion, as completed and completeable only by that which lies beyond it.

If so, then Division Two shows that the analysis of Division One, while lacking nothing, is essentially incomplete, and essentially beyond completion, in a sense that goes beyond the idea that essentially finite human understanding is always capable of further and deeper spirals of articulation. Division Two, rather, suggests that there is something essentially beyond representation in the being whose Being is structured by care, hence something about Dasein that is beyond the grasp of Division One, or of any conceivable supplementation or deepening of the analysis it contains. In effect, the book's internal division returns us to a claim Heidegger makes in its opening pages – 'that in any way of comporting oneself towards entities . . . there lies *a priori* an enigma' (BT, 1: 23). The function of Division Two is thus to disrupt the apparent completeness of Division One, thereby allowing *Being and Time* as a whole to represent the self-concealing and self-disrupting condition of Dasein's Being, and hence of its relation to Being as such. The peculiar way in which Division Two alters nothing and everything in Division

One is thus Heidegger's way of ensuring that *Being and Time* successfully represents Dasein's essentially enigmatic relation to 'the nothing'.

EXCURSUS: HEIDEGGER AND KIERKEGAARD

Heidegger introduced his discussion of death as part of his search for theoretical perspicuity. Human mortality appeared to pose an insuperable obstacle to grasping the ontological structure of human existence as a single, unified whole. But the discussion itself teaches us that a proper understanding of human mortality is also the precondition for any individual human life attaining existential integrity; only by relating to death understood as an impossible possibility can my existence become at once genuinely individual and genuinely whole. In other words, wholeness – properly understood as the unity and integrity belonging to essentially finite, enigmatic beings and their endeavours – has both a theoretical and an existential significance; Being-a-whole is not just the fundamental mark of a good phenomenological analysis, but the touchstone of an authentic relation to death and so to life.

This emphasis upon integrity or wholeness in human existence may appear unmotivated. To be sure, acknowledging one's own mortality must involve acknowledging that death is a threat to existence as such. It thereby highlights that what is at issue in life is not just the content of any given moment but the course of that life. But even if one's life as such is at stake in one's existentiell choices, must one choose in such a way as to make that life into a single, integral whole? Would it not be equally authentic to live a life of multiplicity and diversity, aiming to include as many different activities, achievements and modes of life as possible before death intervenes? Why should the fact that our individual life choices must be seen against the background of the single life of which they are a part mean that we should aim to confer upon it a narrative unity as opposed to a narrative disunity?

Addressing this question properly requires a grasp of Heidegger's account of conscience (the topic of the next two sections), so I will defer delineating his full answer until then. But his seemingly unargued conjunction of the concepts of authenticity, wholeness and death is partly determined by the work of the philosopher with whom these sections on guilt and conscience are implicitly in dialogue – Kierkegaard. For, in effect, Heidegger is offering an alternative answer to a question that Kierkegaard posed, and thereby attempting to distinguish his account of authenticity from the theological competitors with which his idiosyncratic use of ethico-religious concepts, such as guilt and conscience, might seem to align him. Heidegger's proximity to Kierkegaard is thus far more significant than his glancing, critical references to him in the footnotes to sections 40 and 45 would suggest.

Kierkegaard's philosophical pseudonym, Johannes Climacus,[2] shares the Heideggerian view that human beings continuously confront the question of how they should live, and so must locate some standard or value in relation to which that choice might meaningfully be made. Moreover, in so far as that standard is intended to govern every such moment of choice, it confers significance on the whole life that those moments make up – if each choice is made by reference to the same standard, the life which grows from that series of choices will necessarily manifest an underlying unity. Climacus thus presents the question of how best to live as a question about what gives meaning to one's life as a whole, making exactly the conjunction between authenticity and wholeness that Heidegger deploys. In taking over this question in roughly the form in which Climacus poses it, it seems that Heidegger is also taking over his justification for so formulating it.

Climacus goes on to suggest that only a religious answer to the question of life's meaning will do. Suppose that we start by aiming at a specific goal or achievement to give our life meaning – the pursuit of power or wealth, the development of a talent. Since such goals have significance only in so far as the person concerned desires them, what is giving meaning to her life is in reality her wants and dispositions; Climacus calls this the aesthetic form of life. But such dispositions can alter, which means that no such single disposition can confer meaning on my life as a whole: it may change or disappear, but the question remains for as long as I live, so staking my life upon a desire could deprive it of meaning. The only alternative in such circumstances would be to choose another desire upon which to found my life – to aim for power instead of riches, for example; but this would show that the true foundation of my life is not whatever desires I happen to have but my capacity to choose between them.

According to Climacus, then, we can avoid self-deception only by explicitly grounding our lives on our capacity to choose, thus transforming the conditional array of our desires into unconditional values. We might, for example, relate to our sexual impulses by choosing an unconditional commitment to marriage, or choose to view a talent as the basis of a vocation. We thereby choose not to permit changes in these contingencies to alter the shape of our lives, maintaining its unity and integrity regardless of fluctuations in the intensity of our desires, and thereby creating a self for ourselves from ourselves. This is a condensed version of a Kantian, will-based understanding of the ethical form of life; and Climacus's argument for it implies a second reason for connecting authenticity and wholeness. If – as Heidegger suggests – authenticity amounts to establishing and maintaining genuine selfhood, the fluctuations of individual desires and dispositions cannot form an adequate basis for it. The resulting multiplicity of essentially unrelated existential fragments could not cohere into a life that anyone could acknowledge as her own.

Shifting from the aesthetic to the ethical mode of life may, however, be less fundamental than it seems. For the latter understands the human will, the human capacity to hold unconditionally to a choice, as the source of life's meaning; but that capacity is still a part of the person's life, and so a part of that which has to be given meaning as a whole. But no part can give meaning to the whole of which it is a part. With respect to it, as with respect to any of a person's given desires and dispositions, we can still ask: what justifies the choice of the capacity to choose as the basis of one's life? What confers meaning on *it*?

This implies that the question existence sets us is not answerable in terms of anything in that life; life cannot determine its own significance in terms of (some element of) itself. Meaning can only be given to one's life as a whole by relating it to something outside it; for it is only to something outside it that my life can be related *as a whole*. Only such a standard could give a genuinely unconditional answer to the question of the meaning of one's life. Only by relating ourselves to such an absolute Good, and thus relativizing the importance of finite (and so conditional) goods, can we properly answer the question existence poses. And such an absolute Good is, for Climacus, just another name for God; we can relate properly to each moment of our existence only by relating our lives as a whole to God.

Against this background, Heidegger's interpretation of death gains in significance. For, by accepting the Kierkegaardian conjunction between authenticity and wholeness, but arguing that this conjunction can be properly forged by relating appropriately to one's mortality, Heidegger in effect argues that the theological terminus of Climacus's argument is avoidable. By understanding death as one's ownmost possibility and anticipating it in every existential choice one makes, human beings can live authentic and integral lives without having to relate those lives to a transcendent Deity. For, on Heidegger's understanding of human mortality, while a proper grasp of human existence as conditioned does require that one relate it to that which lies beyond its grasp, it does not require that one relate it to some essentially unconditioned thing or being. The relevant horizon is not that of a transcendent Deity, but of nothingness. Kierkegaard is thus right to believe that the question of life's meaning is an inescapable part of human life, and that it can be faced properly only by acknowledging the conditionedness or finitude of that life; but he is wrong to think that acknowledging this finitude requires acknowledging a realm or an entity which lies beyond that finitude. Such talk of a 'beyond' implies that human conditionedness is a limitation rather than a limit, a set of constraints that deprive us of participation in another, better mode of life, rather than a set of conditions that determine the form of any life that is recognizably human. Existential wholeness thus requires only an acknowledgement of human mortality; and only those forms of theological understanding that acknowledge this fact – that understand conditions as limits rather than limitations – are compatible with a proper ontological understanding of human existence.[3]

GUILT AND CONSCIENCE (§§54–60)

Heidegger's reflections on death have so far shown that Dasein's Being-a-whole is ontologically possible, i.e. that this possibility is consonant with the basic structures of Dasein's mode of Being. But it is one thing to demonstrate that it is logically possible for Dasein to individualize itself in an impassioned freedom towards death, and quite another to show that, and how, this possibility can be brought to concrete fruition in the everyday life of a being whose individuality is always already lost in the 'they'. Accordingly, Heidegger next attempts to locate the ontic roots of this ontological possibility – to identify any existentiell testimony to the genuine realizability of Dasein's theoretically posited authenticity.

In its average everyday state of inauthenticity, Dasein is lost to itself. So, for it to achieve authenticity, it must find itself. But it can only begin to do so if it comes to see that it has a self to find, if it overcomes its repression of its potentiality for selfhood. In short, its capacity for authentic individuality must somehow be attested in a way which breaks through its average everyday inauthenticity. Heidegger claims that what bears witness to this possibility for Dasein is the voice of conscience. This existentiell phenomenon is open to, and has been given, a wide variety of interpretations – religious, psychoanalytical, socio-biological. Heidegger neither endorses nor condemns any of these, but rather explores the ontological or existential foundations of the phenomenon to which they refer. His concern is with what makes it possible for Dasein to undergo the experience to which each of these interpretations lays claim. His suggestion is that this experience is the existentiell realization of Dasein's primordial capacity to disclose itself as lost and to call upon itself to attain its ownmost potentiality for selfhood.

As the term 'call' suggests, Heidegger thinks of the voice of conscience as a mode of discourse – a form of communication that attempts to disrupt the idle talk of the they-self to which Dasein is ordinarily attuned, to elicit a responsiveness in Dasein that opposes every aspect of that inauthentic discourse. It must therefore do without hubbub, novelty and ambiguity, and provide no foothold for curiosity. Indeed, if it is transformed into the occasion for endless self-examination or fascinated, narcissistic soliloquies, this voice has been entirely lost, one more victim of the they-self's repressions.

Dasein is its addressee, but its mode of address is not determined by what Dasein counts for in the eyes of others, what its public role and value may be, nor by what it may have taken up as the right way to live its life. It addresses Dasein purely as a being whose Being is in each case mine, i.e. for whom genuine individuality is a possibility. Accordingly, its call is devoid of content: it asserts nothing, gives no information about world events, and no blueprints for living – it merely summons Dasein before itself, holding up every facet of its existence, each aspect of its life choices, for trial before its capacity to be itself. It calls Dasein forth to its ownmost possibilities, without venturing to dictate what those possibilities might or should be; for any such dictation could only further repress Dasein's capacity to take over its own life. In short, 'conscience discourses solely and constantly in the mode of keeping silent' (BT, 56: 318).

Who, then, addresses Dasein in this way? Whose is the voice of conscience? We cannot specify the caller's concrete features, for it has no identity other than as the one who calls; the summoner exists only as that which summons Dasein to itself. But this voice is one that Dasein hears within itself, and is usually understood as an aspect of Dasein itself; so can we not conclude that, in the voice of conscience, Dasein calls to itself? For Heidegger, matters are more complex. He agrees that the voice of conscience is not the voice of someone *other* than the Dasein to whom the call is addressed, not the voice of a third party. But neither are Dasein-as-addressee and Dasein-as-addresser one and the same. For the Dasein to whom appeal is made is lost in the 'they', whereas the Dasein who makes the appeal is not (and could not be, if its silent voice is to disrupt the discourse of the they-self). After all, on Heidegger's account, part of Dasein's lostness in the they-self is its being lost to any conception of itself as lost, as possessed of a capacity for authentic individuality. This fits our everyday experience of conscience as a voice that speaks against our expectations and even against our will: its demands are ones to which we have no plans or desire to accede. But, then, the voice of conscience both is and is not the voice of the

Dasein to whom it speaks – 'the call comes *from* me and yet *from beyond me*' (BT, 57: 320). How are we to make sense of Dasein's passivity in relation to this voice? How can its being the voice of Dasein be reconciled with the fact that it is characteristically experienced as a call made upon rather than by Dasein?

This passive aspect of the voice of conscience suggests that it relates to Dasein's thrownness – that the voice of conscience is somehow expressive of the fact that Dasein is always already delivered over to the task of existing, placed in a particular situation that it did not choose to occupy, but from which it must nevertheless choose how to go on with its life. This is Dasein's fundamental uncanniness: the state in which it finds itself is never all that it is or could be, and so never something with which it can fully identify or to which it can be reduced – so that Dasein can never regard itself as domesticated, fully at-home with whatever state or form of life and world it finds itself inhabiting. It is from this thrownness into existential responsibility that the they-self flees; but the voice of conscience recalls Dasein to this fact about itself, and thereby throws the individual into an anxious confrontation with its own potentiality for genuine individuality. In short, the voice of conscience is that of Dasein in so far as it 'finds itself in the very depths of its uncanniness' (BT, 57: 321).

This is why the one who calls through the voice of conscience is definable by nothing more concrete than the fact of its calling: it is the voice of Dasein as 'not-at-home', as the bare there-Being (Da-sein) in the nothingness which remains when it is wrenched from its familiar absorption in the world, and that world stands forth as the arena for Dasein's projective understanding. Nothing could be more alien to the they-self than the self that confronts its potentiality for authentic existence; nothing is more likely to be experienced by the they-self as at once within and without the self. And, since the voice of conscience is the voice of Dasein as thrown projection, the voice which summons it from its lostness to confront its inescapably personal abandonment to the task of existing, it can be thought of as the call of care. In other words, the call of conscience is ontologically possible only because the very basis of Dasein's Being is care.

This is Heidegger's ontological explanation for the ontical fact that the voice of conscience is often heard as accusing us, as identifying the one it addresses as being guilty. Conceptually, guilt is connected with indebtedness and responsibility. A guilty person is responsible for atoning for herself, making reparation for some deprivation or lack that she has inflicted on others, which in turn presupposes that she herself is lacking in something – that she has been, and is, deficient in some way, and is responsible for that deficiency. In short, being guilty is a matter of being responsible for, being the basis of, a nullity. But then the ontic phenomenon of guilt reflects the fundamental ontological structure of Dasein's existence as thrown projection.

Through existing, Dasein realizes one of the existentiell possibilities that its situation determines as available to it; it acts *on the basis* of the particular state of self and world in which it finds itself. But, of course, it never has complete control over that state and the restrictions it imposes; the capacity for projective commitment must always be deployed from within some particular context or horizon, and so could never wholly determine its structure:

> In being a basis – that is, in existing as thrown – Dasein constantly lags behind its possibilities. It is never existent *before* its basis, but only *from it* and as *this basis.* Thus 'Being-a-basis' means *never* to have power over one's ownmost being from the ground up. This 'not' belongs to the existential meaning of 'thrownness'.
>
> (BT, 58: 330)

However, nullity is integral to Dasein's capacity for projection as well as to its thrownness. For, in projecting upon one particular possibility, Dasein thereby negates all other possibilities: the realization of any existentiell choice is the non-realization of all others. 'Thus, "care" – Dasein's Being – means, as thrown projection, Being-the-basis of a nullity (and this Being-the-basis is itself null)' (BT, 58: 331). In short, human existence as such amounts to the null Being-the-basis of a nullity; Dasein as such is guilty.

The authenticity to which conscience calls Dasein is thus not an existentiell mode in which Dasein would no longer be guilty.

Excuses or acts of reparation and reform might eradicate the ontic guilt of a specific action, but ontological guilt, being a condition of human existence, is originary and ineradicable. Authenticity, rather, demands that one project upon one's ownmost potentiality for being guilty. The aim is not to overcome or transcend guilt, since that would amount to transcending one's thrownness; it means taking responsibility for the particular basis into which one is thrown and the particular projections one makes upon that basis, to make one's necessarily guilty existence one's own rather than that of the they-self. A readiness to take on responsibility in this way, to be indebted to oneself, amounts to a willingness to be appealed to by the voice of conscience – a readiness to make existential decisions in the light of one's ownmost, authentic potentiality for Being-guilty. It amounts, in short, to choosing to have a conscience as opposed to repressing it. The response for which the voice of conscience is seeking is thus not the adoption of some particular schedule of moral rights and wrongs, some specific calculus of debt and credit. The response it seeks is responsiveness, the desire to have a conscience. To cultivate such a desire is to put oneself in servitude to one's capacity for individuality; it is to choose oneself.

Since wanting to have a conscience amounts to Dasein's projecting upon its ownmost potentiality for Being-guilty, we can think of it as a mode of understanding. But, in the tripartite care-structure of Dasein's Being, to every mode of understanding a particular state-of-mind and a particular mode of discourse belong. We saw that the announcement of Dasein's uncanniness elicits anxiety; and, as the indefiniteness of the call conscience makes and the response it demands makes clear, the mode of discourse which corresponds to this anxiety is one of keeping silent, of reticence. The particular form of self-disclosedness that the voice of conscience elicits in Dasein is thus a reticent self-projection upon one's ownmost Being-guilty in which one is ready for anxiety. Heidegger labels it 'resoluteness'.

As a mode of Being-in-the-world, resoluteness does not isolate Dasein or detach it entirely from its world. Rather, it returns Dasein to its particular place in its world, to its specific concernful relations with entities and solicitous relations with others, in order to discover what its possibilities in that situation really are and to seize upon them in whatever way is most genuinely its own. Resoluteness is therefore inherently indefinite: if the concrete disclosures and projections which make it up must be responsive to the particularity of its context, then no existentiell blueprints for authenticity can arise from a fundamental ontology. In fact, it is only through the disclosive understanding of a concrete act of resolution that a particular context – hitherto volatilized by the ambiguity, curiosity and novelty-hunger of the they-self – is given existential definition at all. The constitution of Dasein's place in the world as a locus of authentic existentiell choice – as what Heidegger calls a 'situation' – is thus not something resoluteness presupposes, but rather something it brings about. To be resolute involves not simply projecting upon whichever existential possibility from a given range is most authentically one's own, but projecting one's context as possessed of a definite range of existential possibilities in the first place. Resoluteness constitutes the context of its own activity.

THE ATTESTATION OF *BEING AND TIME*

It seems, then, that Heidegger can marry the various components of his analysis of Dasein into a coherent whole. His various characterizations of human existence as thrown projection, care, Being-towards-death and Being-guilty dovetail rather than conflict with one another. They are complementary specifications of the same ontological structure from differing depths and angles of analysis. But one of his declared goals in this particular chapter remains unfulfilled.

For his account of conscience is supposed to provide some existentiell proof that a being typically mired in inauthenticity might nonetheless attain authenticity. In one sense, of course, it does just that: if the account is accurate, then that voice articulates the call of Dasein's uncanniness, and so constitutes a trace within everyday existentiell inauthenticity of that aspect of Dasein which is anxious about its ownmost potentiality for authentic existence. But, for Heidegger, the voice utters a call that Dasein makes from itself to itself; it is the voice of Dasein's repressed but not extinguished capacity for genuine selfhood. And, yet, if that capacity is genuinely repressed, how can it possibly speak out? If it can, its repression must already have been lifted; but it is just that lifting, that transition from inauthenticity to authenticity, which the call of conscience is supposedly invoked to explain.

The central difficulty is that Heidegger conceives of Dasein as inherently split or doubled.[4] All human beings are capable of living authentically or inauthentically: either they are lost in the distractions of the they-self (while retaining the capacity for wrenching themselves away from it), or they have realized the existentiell possibilities that give expression to their real individuality (while remaining vulnerable to a falling back into loss of self). The transition from inauthentic to authentic existence therefore involves a shift in the internal economy of these dual-aspect beings: the capacity for genuine individuality must come to eclipse the capacity for non-individuality which has hitherto eclipsed it. But Heidegger conceives of this transition as brought about by Dasein's own resources – 'the call undoubtedly does not come from someone else who is with me in the world' (BT, 57: 320) – and such a vision of the self-overcoming of self-imposed darkness is difficult to render coherent. Heidegger claims that the transition is brought about by the very aspect of the self that benefits from it – by its eclipsed capacity for authenticity: '[Dasein's] ownmost potentiality-for-Being-its-Self functions as the caller' (BT, 57: 320). But this amounts to claiming that a capacity in eclipse can bring about its own emergence from eclipse. The only available alternative explanation is that the capacity at present eclipsing the self's capacity for authenticity might place itself in eclipse – which seems no less incoherent. In short, the transition with which Heidegger is concerned seems inexplicable in his own terms.

The difficulty is fundamental and, I believe, insuperable without some modification of the model Heidegger has offered. But there is one obvious modification that might solve the difficulty while preserving the basic outlines of his understanding of conscience: we can drop the claim that the call of conscience does not come from someone else who is with us in the world. What if we claimed instead that the call of conscience is in fact articulated by a third party, by someone else who diagnoses us as lost in the they-self, and has an interest in our overcoming that inauthenticity and freeing our capacity to live a genuinely individual life? The intervention of such a person would constitute an external disruption of the hermetic, self-reinforcing dispersal of Dasein in the they-self, a way of recalling the self to its own possibilities without requiring an incoherent process of internal bootstrapping. She would, in a sense, be speaking from outside or beyond us; but Heidegger has stressed that a perceived externality is one characteristic of the voice of conscience. Moreover, if this person's aim is to help us recover our capacity for selfhood, our autonomy, she could not consistently wish to impose upon us a specific blueprint for living or in any other way substitute a form of servitude to herself for our present servitude to the 'they'. In fact, her only aim would, rather, be that of recalling us to the fact of our capacity for individuality, and urging us to listen to the specific demands it makes upon us. In so doing, she would function as an external representative of an aspect of ourselves, her voice going proxy for the call of our ownmost potentiality for authenticity, a call that has at present been repressed but which nonetheless constitutes our innermost self; in that sense, her voice would be speaking from within us.

In short, the voice of a third party, whose reticent appeal acknowledged the logic I have just outlined, would be perceived by us as possessing just the phenomenal characteristics Heidegger uses to define the voice of conscience: 'The call comes *from* me and yet *from beyond* me' (BT, 57: 320). It then seems significant that, when Heidegger briefly refers to the voice of conscience in his discussion of language, he talks of 'hearing the voice of the friend whom every Dasein carries with it' (BT, 34: 206);[5] and that he should note in passing that 'Dasein . . . can become the conscience of Others' (BT, 60: 344).

If, however, inauthentic Dasein is incapable of uttering the call of conscience, how can it be capable of hearing that call when it is made by another? If part of Dasein's lostness in the they-self is its loss of any conception of itself as lost, as capable of anything other than its present state, how could the friend's call to recognize that its present state is inauthentic (and hence alterable) actually penetrate its repression of any such awareness? If it could, then surely its addressee must already in part have made the very transition that the reception of the call is supposed to explain. Clearly, then, if the friend is to be heard, she must create the conditions for her own audibility. But how?

Inauthentic Dasein's selfhood is lost in the they-self; ontologically speaking, there is no self–other differentiation in the 'they', and so no internal self-differentiation in its members – lacking any conception of being other than it is, Dasein conflates its existential potential and its existentiell actuality, and represses its uncanniness. When, however, Dasein encounters an authentic friend, her mode of existence disrupts the undifferentiated mass of the 'they'; her selfhood is not lost in a slavish identification with (or a slavish differentiation from) others, so she cannot confirm Dasein in its anonymity by mirroring it, and she prevents Dasein from relating inauthentically to her. For Dasein could mirror another who exists as separate and self-determining, and who relates to others as genuinely other, only by relating to her as other and to itself as other to that other, i.e. as a separate, self-determining individual. This amounts to Dasein acknowledging the mineness of its existence, and so its internal self-differentiation (the uncanny non-coincidence of what it is and what it might be). In short, an encounter with a genuine other disrupts Dasein's lostness by awakening otherness in Dasein itself; Dasein's relation to that other instantiates a mode of its possible self-relation (a relation to itself as other, as not self-identical). Put otherwise, it induces an anxious realization of itself as a separate, self-responsible being with a life that it must lead, and so of its existence as its own, non-relational and not-to-be-outstripped. This amounts to an anxious acknowledgement of its mortality, the anticipatory state that Heidegger earlier defined as the existentiell pivot from self-dispersal to self-constancy. This is how the sheer fact of the friend's existence creates in those to whom she relates herself the conditions for the audibility of her call to individuality.

This leaves one final problem: if Dasein's transformation to authenticity presupposes an authentic friend, how did the friend achieve authenticity? Does not our 'solution' to Dasein's boot-strapping problem simply displace it on to this third party, and so leave us no further forward? This important question is one that can only be addressed using the material examined in Chapter 7, so we must defer its resolution until then. What I can spell out here, however, is the reflexive potential of this modified version of the Heideggerian model of conscience – its applicability as a model for understanding the role of the text in which it is developed.

For, of course, Heidegger's conception of Dasein as split, with its capacity for authenticity eclipsing or being eclipsed by its capacity for inauthenticity, is intended to apply to his readers. As students of philosophy, they will be immersed in the prevailing modes of that discipline; and since philosophizing is a mode of Dasein's Being, its everyday enactments will be as imbued with inauthenticity as will those of any human activity. In short, Heidegger conceives of the readers of *Being and Time* as inauthentic, although capable of authenticity. Since, however, outlining an insightful fundamental ontology of Dasein would necessarily be an achievement of authentic philosophizing, and since that is exactly what *Being and Time* claims to develop, Heidegger must regard the author of *Being and Time* – himself – as having achieved an authentic mode of human existence (while not being immune to the temptations of inauthenticity). Add to this the fact that providing such a fundamental ontology to his readers amounts to an attempt to facilitate their transition from inauthentic to authentic philosophizing, and we have a picture of Heidegger's relations to his readers that precisely matches the modified model of conscience I just introduced.

Heidegger appears as the voice of conscience in philosophy, offering himself as an impersonal representative of the capacity for authentic thinking that exists in every one of his readers, presenting them not with blueprints for living but with a portrait of themselves as mired in inauthenticity, in order to recall them to knowledge of themselves as capable of authentic thought, and thereby to encourage them to overcome their repression of that capacity and to think for themselves. In short, Heidegger's words offer themselves as a pivot for their readers' self-transformation, as at once a mirror in which their present inauthenticity is reflected back to them and as a medium through which they might attain authenticity.

Why, then, should Heidegger emphatically exclude the possibility of our modified model of the voice of conscience by declaring that it can never be the voice of an actual other, a third party? One possible answer is that he is attempting to preserve the idea that the transformation from inauthenticity to authenticity can be brought about through the relevant individual's own resources – that Dasein can originate its own rebirth. But, of course, in claiming the capacity to present a fundamental ontology of Dasein (of which this analysis of conscience forms a part), the author of *Being and Time* lays claim to a position of authenticity as a philosopher, and so implicitly identifies himself as having managed the transition from an inauthentic to an authentic mode of existence. His unmodified model of conscience allows him to present himself as having done so entirely out of his own resources, as having single-handedly created his fundamental ontology and his deconstruction of the philosophical tradition he inherited. His achievement appears as solely and exclusively his, as if it had sprung fully formed from his own forehead. In particular, it provides a subliminal justification of his otherwise puzzling decision to repress entirely the role that his teacher, Husserl, played in the origination of his own thinking and his own investigations – to repress the voice of conscience that Husserl clearly represented for him.

Of course, such a mode of self-presentation makes it difficult for Heidegger to acknowledge that his model of conscience can also account for the relation in which he stands to his readers, that the voice of his text is the voice of conscience, the call of care – for how can he explicitly declare that, while others require the intervention of his voice to reactivate their potentiality for authenticity, he alone stood in no such need, that he benefited from no one in the way his readers will benefit from him? And what this shows, I believe, is the frightening depth of Heidegger's need to think of himself as self-originating. It is not necessarily a constant need, or at least one that constantly overwhelmed him; indeed, as Chapter 7 of this book will argue, other stretches of his text implicitly deny that his ideas are entirely self-originated. But, at this point, it is difficult to avoid the conclusion that Heidegger's need to deny his own dependence upon others has led to a fundamental mutilation of the potential wholeness and integrity of his text – a distortion of the fit between its form and its content that amounts to a distortion of its authenticity.

But I want to end this long and complex chapter by underlining a respect in which the form and the context of this text do achieve a genuinely authentic fit. To see this, we first need to recall the extent to which Heidegger's analysis of conscience and guilt confirms the implication of his analysis of death – namely, that Dasein is internally related in its Being to nothingness, nullity and negation. To say that Dasein is Being-guilty just *is* to say that it is the null Being-the-basis of a nullity, and hence that something about the ground of our projections will always exceed our comprehending grasp; and the voice of conscience is Dasein discoursing to itself in the mode of keeping silent – that is, it reveals the being of discourse as paradigmatically manifest in saying nothing, or rather in a dimension of significance that goes beyond the specifiable content of a speech act. For this silent voice does not demand that anything specific happen in the world, and so nothing specific could constitute its satisfaction. More precisely, beyond any specific existentiell demands we interpret it as making, the voice of conscience always makes the further demand that we regard our subjection to demand as such as unredeemable through the satisfaction of those specific demands.

What the voice of conscience speaks against, therefore, is our inveterate tendency to conflate our existential potential with our existentiell actuality; so what it silently opens up is Dasein's internal otherness, its relation to itself as other, as not self-identical but rather transitional or self-transcending. And this implies that inauthenticity is a matter of Dasein's enacting an understanding of itself as essentially self-identical, as capable of coinciding with itself and fulfilling its nature. But, if Heidegger means his text to be the voice of conscience for his readers, then, in order to meet the standards that its own analyses set, it must at all costs avoid coinciding with itself. Can it be so understood? It can if we interpret the apparent completeness and self-sufficiency of Division One as the text's enactment of exactly the inauthentic absorption in specific work-environments (the self's untroubled identification with its world) and the undifferentiatedness of the they-self (the self's untroubled coincidence with Others and hence with itself) that it identifies as signals of average everyday concern and solicitude. On this interpretation, it is the internal differentiation of *Being and Time* between Divisions One and Two that grounds its overall claim to be providing an authentic existential analytic of Dasein, and hence a way of turning its readers from inauthenticity to authenticity as philosophers and as individuals. It is Division Two's refusal to coincide with Division One – its refusal to accept that its predecessor's characterization of the care-structure is complete and self-sufficient, simply coinciding with the Being of the being under analysis – that gives *Being and Time* its authentic unity: the book's internal self-transcendence or self-negation is its way of Being-a-textual-whole. For the irruptive advent of Division Two – at once unfolding from certain specific aspects of the analysis of Division One (involving angst and Being-ahead-of-oneself) and entirely reorienting every aspect of it – enacts the way in which an authentic self-understanding is to be wrenched from the inauthentic grasp of ourselves with which the book tells us we will always already begin, both as individual Dasein and as philosophers. Hence, an authentic grasp of Heidegger's existential analytic depends upon seeing it as deliberately, unavoidably, disrupting itself from within (by striving to represent Dasein's internal relation to what is beyond representation), and thereby aiming to achieve the non-self-coincidence that is the mark of anxious, anticipatory resoluteness.

NOTES

1   See L. Wittgenstein, *Tractatus Logico-Philosophicus* (London: Routledge and Kegan Paul, 1922), 6.4311ff.

2   See S. Kierkegaard, *Concluding Unscientific Postscript*, trans. H. V. and E. H. Hong (Princeton, NJ: Princeton University Press, 1992). The question of the significance of Kierkegaard's use of pseudonyms is controversial, and particularly so in the case of this book; for safety's sake, I will attribute the views expressed in it to its pseudonymous author.

3   Whether Heidegger is right to think that Climacus's account of what it is to relate human finitude to the Absolute falls into the trap of

misinterpreting human conditionedness is a moot point. See my *Faith and Reason* (London: Duckworth, 1994) for an argument that Climacus is not guilty as charged; see M. Weston, *Kierkegaard and Modern Continental Philosophy* (London: Routledge, 1994) for a Kierkegaardian critique of Heidegger.

4   In articulating this difficulty, coming to see its significance and attempting to develop a way of accommodating it that is not wholly alien to Heidegger's self-conception, I am drawing upon a specific set of terms and a general conception of the philosophical enterprise developed in the work of Stanley Cavell: see in particular his Carus lectures, *Conditions Handsome and Unhandsome* (Chicago: University of Chicago Press, 1990). In so doing, I hope to convince the reader that the perfectionist model of philosophical writing that Cavell claims to find at work in the texts of Emerson, Thoreau and Wittgenstein (among others) can also be seen to control the early Heidegger's conception of his endeavours.

5   Derrida makes much of this point in his essay: 'Heidegger's ear: Philopolemology', in J. Sallis (ed.) *Reading Heidegger: Commemorations* (Bloomington, Ind.: Indiana University Press, 1994).

6

HEIDEGGER'S (RE)VISIONARY MOMENT: TIME AS THE HUMAN HORIZON

(Being and Time, §§61–71)

Our brief discussion of the friend as the voice of conscience implied a connection between Dasein's willingness to attend to that voice and its anticipation of its death. In the sections to be examined next, Heidegger argues that these two elements of Dasein's authenticity are simply different facets of one and the same mode of existence. This prepares the ground for outlining the ontological preconditions of Dasein's Being as care, thereby definitively establishing an internal relation between the Being of Dasein and time. In so doing, Heidegger explicitly develops two other themes also highlighted at the end of the previous chapter: first, that to understand Dasein's Being is to understand another aspect of its internal relation to the nothing; and, second, that the conclusions established in his text control the ways in which that text is written and should be read, hence that the content and the form of authentic philosophical writing must be properly related to one another.

MORTALITY AND NULLITY: THE FORM OF HUMAN FINITUDE (§§61–2)

The connection between anticipation and resolution depends on the internal relation between Heidegger's dual characterization of Dasein's Being as Being-towards-death and as Being-guilty (Being-the-null-basis of a nullity); for both characterizations invoke different inflections of a single conception of negativity at the heart of human existence. Together, they entail that human beings properly understand the significance of their existentiell choices only if they make them knowing that each such moment of decision might be their last, and that each constitutes a situation into which they were thrown and from which they must project themselves.

These are simply two interrelated marks of the conditionedness or finitude of human existence – finitude as mortality and finitude as nullity: they envision each moment of human existence as shadowed by the possibility of its own impossibility, by the absence of total control over its own antecedents, and by the negation of competing but unrealized possibilities. Accordingly, human beings cannot authentically confront their concrete moments of existential choice unless they grasp the full complexity or depth of their finitude. They cannot resolutely confront them as the null basis of a nullity without acknowledging the possibility of their utter nullification (i.e. without anticipating death); and they cannot properly anticipate their own mortality without confronting their choice-situations as themselves doubly marked by death – the death of the preceding moment (no longer alterable but forever determinative) and the death of their other unrealized possibilities (no longer actualizable but forever what-might-have-been). In short, the only authentic mode of resoluteness is anticipatory resoluteness: the only authentic mode of anticipation is resolute anticipation.

The desired impact of the voice of conscience on an attentive Dasein confirms that anticipation is the authentic existentiell modification of resoluteness. That voice wrenches Dasein away from its lostness in the 'they' and returns it to its ownmost potentiality for selfhood. It individualizes Dasein, forcing it to confront its underlying non-relationality; and it recalls Dasein to a conception of its own existence as essentially and inescapably Being-guilty. The resoluteness it calls for involves establishing and maintaining constancy with respect to the real lineaments of Dasein's situation, but avoiding the a priori imposition of specific blueprints for living. But the particular mode of existence that best answers to these very precise demands – the mode of projection that best responds to the voice of conscience – would be Dasein's ownmost, non-relational, not-to-be-outstripped, certain and yet indefinite possibility; and that is simply a description of Being-towards-death. In other words, 'resoluteness is authentically and wholly what it can be, only as *anticipatory resoluteness*' (BT, 62: 356).

It follows that anticipatory resoluteness will give any Dasein capable of achieving it the only species of unity or wholeness attainable by a being with its distinctively existential mode of Being. Here, Heidegger's analysis explicitly touches on and supplements Kierkegaard's reasons for connecting authenticity with wholeness. For any human being whose resolute grasp of her choice-situation involves projecting herself upon a given possibility against a background awareness of her own mortality will view the relevant moment not simply as if it were her last, but also as a particular, non-repeatable moment in the wider context of her life. Seen in terms of her own possible impossibility, any given moment in a person's existence is revealed not just as utterly contingent in itself, but as part of an utterly contingent life – one with a very specific origin and history, one which will end at a specific point in a specific way, a sequence which might have been different but whose particularity is now the horizon within which she must either attain or fail to attain true individuality. But individuality is not just a matter of making decision after decision, each of which is genuinely expressive of herself rather than of the 'they'; it means leading a life that is genuinely her own.

Accordingly, placing any particular moment of decision within the context of a single and singular life must be the goal of any genuine act of resolution. Resolutely grasping one's existential responsibilities means disclosing the true lineaments of one's decision-making context, determining it as a situation for existentiell choice; and that is a matter of contextualizing it, of properly grasping the moment as emerging from the constraints and free-doms of the preceding moment and as providing a basis for projecting upon the available possibilities of the coming moment. But fully comprehending the specificity of that moment would involve placing it in a context wider than the immediate past and future. It would mean seeing it as the point to which one's life has led, and from which the remainder of one's life will acquire a specific orientation.

Such a contextualization must of course acknowledge that one's life cannot be grasped as a whole in any absolute or unconditional sense; for it must be grasped by the being whose life it is, and so from a point within it rather than from some fantasized point out-side it, which means that Dasein's comprehending grasp of itself will necessarily encounter constitutive limits, reflecting the fact that its Being is the null basis of a nullity. Nor does such contextual-ization require that one's life as a whole should have a single, overarching plot – with everything in it subordinate to a single goal; narrative unity need not be monomaniacal. But resolute anticipa-tion would require avoiding the complete fragmentation implicit in the Kierkegaardian portrait of the aesthetic life; it would require continually striving to understand the twists and turns of one's life as episodes in a single story. Relating oneself to all moments of decision in this way would, accordingly, mean viewing every moment as one in which the significance of one's life as a whole is at stake; and that simply reformulates Heidegger's conception of living in the full awareness of one's mortality. So, by actualizing its potential for Being-a-whole, Dasein would enact an authentic mode of Being-towards-death.

PHILOSOPHICAL INTEGRITY AND AUTHENTICITY (§§62–4)

At this point, however, Heidegger acknowledges a significant shift in the focus of his investigation:

> *The question of the potentiality-for-Being-a-whole is one which is factical and existentiell. It is answered by Dasein as resolute.* The question of Dasein's potentiality-for-Being-a-whole has now fully sloughed off the

> character indicated at the beginning, when we treated it as if it were just a theoretical or methodological question of the analytic of Dasein, arising from the endeavour to have the whole of Dasein completely 'given'. The question of Dasein's totality, which at the beginning we discussed only with regard to ontological method, has its justification, but only because the ground for that justification goes back to an ontical possibility of Dasein.
>
> (BT, 62: 357)

Attaining a perspective upon Dasein as a totality or whole originally appeared as a methodological imperative: Heidegger's overt concern was to demonstrate that the seemingly disparate elements of his analysis of Being-in-the-world in fact formed an articulated whole, that his ontological analysis was a comprehensive, integrated and surveyable treatment of the human way of being. Now, we are told that its covert inspiration lies in its relation to an ontical possibility of Dasein: Heidegger's supposedly impersonal methodological interest in wholeness is in reality a personal interest in a particular existentiell possibility – attaining anticipatory resoluteness.

He thereby acknowledges one implication of the generally reflexive nature of his enterprise. For, of course, Heidegger is a human being writing an analytical account of the underlying structures of the human way of being; so every element of that analysis must apply to himself, and in particular to his way of engaging in philosophical analysis and composing philosophical prose. But a key insight of that analysis is that the human way of being is grounded in care; and the care-structure has a very specific character:

> Because it is primordially constituted by care, any Dasein is already ahead of itself. As Being, it has already projected itself upon definite possibilities of its existence; and in such existentiell projections, it has, in a pre-ontological manner, also projected something like existence and Being. *Like all research*, the research that wants to develop and conceptualize that kind of Being that belongs to existence is *itself a kind of Being which disclosive Dasein possesses*; can such research be denied this projecting which is essential to Dasein?
>
> (BT, 63: 363)

The ontological investigation of which *Being and Time* is a record is itself a mode of Dasein's Being, an enactment by a human individual of one existentiell possibility. It must therefore be guided by a fore-conception of that Being; and, as the realization of a possibility by a given individual, it must involve that individual projecting upon a particular existentiell option. Heidegger's confession identifies the particular existentiell option he aims to realize as that of anticipatory resoluteness, Being-a-whole. In other words, he is projecting upon the specific ontic possibility of authentic Being-in-the-world, and his writings are an essential component of that projection. The seemingly impersonal philosophical activity of which *Being and Time* is the articulate record is in fact part of Heidegger's attempt to make his own life an integral and singular whole – the life of an authentic individual. And, as we have seen, the only alternative to a philosopher's grounding her activity upon an authentic existentiell possibility is her grounding it upon an inauthentic one. In short, since a philosopher is a human being whose life is necessarily structured by the projective understanding of care, her practice and her conclusions cannot transcend or avoid the question of personal authenticity.

So much for professional philosophical detachment. For Heidegger, the very idea is an illusion rooted in Dasein's average everyday repression of its capacity for authenticity, and in philosophy's average everyday repression of its knowledge that – with respect to investigations of human ontology – the investigator is also that which is investigated. In this respect, Kant stands as exemplary. His understanding of the selfhood of human beings avoids the obvious modes of inauthentic human self-understanding. He opposes the Cartesian conception of the human subject as a present-at-hand thinking substance with his claim that the 'I think' represents a purely formal unity, the transcendental unity of apperception (the relatedness of all subjective representations in and to one consciousness). But he conceives of those representations as empirical phenomena constantly present to the 'I' while the 'I' is constantly present to them, and so models their mutual relatedness in terms entirely inappropriate to an entity with the Being of Dasein. While dimly perceiving the inherent directedness of human perceptions –

their necessarily being perceptions of something, subjective perceptions of an objective world – he fails to follow up this glimpse of Dasein's inherently worldly existence because his model of that directedness derives from a particular mode of the being of objects. And it is the distinguishing characteristic of inauthentic Dasein to interpret itself in just those terms; they are the handiest available to a creature that has fallen into its world, immersing itself in the objects which thereafter absorb it. If even so great a philosopher as Kant cannot struggle free of such misconceptions, then the inauthenticity of average everyday philosophizing must be as pervasive and deep-rooted as in any other human activity.

But Heidegger's general diagnosis of philosophers, as systematically denying both the fact and the nature of their own humanity, is not purely a manifestation of his own personal attempt to overcome that professional deformity. The authentic ontology of Dasein recounted in *Being and Time* is not presented to his fellow-philosophers purely to confirm his own authenticity (although it inevitably attests to precisely that). It is also designed to disrupt the inauthentic self-understandings and modes of existence of its readers, to remind them that they too are capable of authenticity, and thereby to serve as a fulcrum upon which they might shift their own lives from lostness to reorientation, from constancy to the not-self of the 'they' to constancy to themselves and to a life that is genuinely their own.

If, as readers, we fail to acknowledge Heidegger's conception of his relation to us, then in effect we simply continue to flee from the voice of conscience and its demand for resoluteness. For authentic resoluteness must grasp the true lineaments of every moment of life, understood as a situation for existentiell choice; and sitting for a certain number of hours reading *Being and Time* is itself such a choice – a particular way of enacting one's existence, and one which places us in a certain field of existentiell possibilities to which we can relate either authentically or inauthentically. Studying philosophy is not an alternative to existing but a mode of existing; and, when it takes the form of studying a philosophical text, doing so authentically must involve acknowledging the fact that the words we are reading were chosen and ordered by another human being, and that our reading those words is not an accident or a necessity but a specific choice that we have made. To pass over the fact that even philosophy books are written by human beings to be read by human beings amounts to repressing the knowledge that studying this philosophical text is a mode of existing, a choice to spend one's time in a particular way with a particular other; so it amounts to denying one's own humanity – denying the fact that even readers and writers of philosophy are human beings.

THE TEMPORALITY OF CARE: THROWN PROJECTION (§§65–8)

The full significance of both the existentiell and the ontological aspects of Heidegger's analysis of Being-a-whole depends upon a further step in that analysis – laying bare the underlying ontological meaning of Dasein's Being as care.

Heidegger thinks of this step as articulating the *meaning* of Dasein's Being as care, where 'meaning' signifies 'the upon-which of a primary projection in terms of which something can be conceived in its possibility as that which it is' (BT, 65: 371). In effect, then, he is exploring the conditions for the possibility of the articulated structural whole that is care. Anticipatory resoluteness, being a mode of human existence, must be an inflection of the care-structure; so any fundamental ontological presuppositions pertaining to authentic resoluteness must also be fundamental to the care-structure. They will, in effect, provide an indirect route to Heidegger's primary goal.

It quickly becomes evident that authentic resoluteness presupposes Dasein's openness to time. It transforms Dasein's potential for authenticity into actuality – a transformation that is inevitably oriented towards the future, towards a future state of the self that Dasein will (and wills to) be. Such authentic projection requires grasping Dasein as the basis for that projection, which means grasping it as null – as essentially Being-guilty. But that is a matter of Dasein's acknowledging itself as it has already been, acknowledging its past as an ineradicable part of its present existence. And, since resoluteness discloses the current moment of Dasein's existence as a situation for choice and action, it also presupposes Dasein's openness to the present – its capacity to let itself be encountered by that which is present to it in its existential context (its 'there'). Resoluteness thus implies a triple but internally related openness to future, past and present. No single openness could exist without the others, but, in so far as resoluteness is anticipatory, a certain priority for Dasein's openness to the future is implied. The limitations, determinations and opportunities bestowed by past and present are grasped so that Dasein might project itself upon its ownmost existentiell possibilities, might open itself to that which is most truly itself as it comes towards it from the future:

> Coming back to itself futurally, resoluteness brings itself into the Situation by making present. The character of 'having been' arises from the future, and in such a way that the future which 'has been' (or better, which 'is in the process of having been') releases from itself the Present. This phenomenon has the unity of a future which makes present in the process of having been: we designate it as *temporality*.
>
> (BT, 65: 374)

In other words, temporality is the meaning of care – the basis of the primordial unity of the care-structure. That totality was previously defined as ahead-of-itself-already-Being-in (a world) as Being-alongside (entities encountered within the world); it reflects Dasein's existence as thrown projection, living a moment that is grounded in previous moments and that in turn grounds moments to come, and so implicitly presupposes openness to time. 'Ahead-of-itself' presupposes Dasein's openness to the future; 'already-Being-in' indicates its openness to the past; and 'Being-alongside' alludes to the process of making present. Once again, the three aspects of temporal openness are internally related, but their ordering in Heidegger's definition registers the relative priority of futurity, which reflects the fundamental ontological fact that existence is a matter of projecting thrownness through present action. Just as resoluteness finds its authentic flowering in anticipation, so the primary meaning of existentiality is the future.

Heidegger's conclusion, therefore, is that the meaning or under-lying significance of the Being of Dasein is temporality. It is what makes possible the unity of existence, facticity and falling to which the tripartite structure of care alludes. We have finally arrived at the theme registered in the title of his book. If Dasein's capacity to relate itself to Being (its own and that of any other being) is of its essence, and if that essence is grounded in its relation to time, then any proper answer to the question of the meaning of Being will inevitably relate Being to time. But what that relation might signify depends upon what Heidegger means by 'time'; and his provisional understanding of the term is far from orthodox.

First, since temporality is the meaning of the Being of Dasein, it cannot be a medium or framework to which Dasein is merely exter-nally or contingently related, something whose essence is entirely independent of Dasein. Heidegger's idea is not that human beings necessarily exist *in* time, but rather that they exist *as* temporality, that human existence most fundamentally *is* temporality. Second, since the care-structure is an articulated unity, the same must be true of that which makes it possible: in other words temporality does not consist of three logically or metaphysically distinct dimen-sions or elements, but is an essentially integral phenomenon. Third, the terminological shift from talk of 'time' to talk of 'temporality', from what sounds like the label for a thing to a term that connotes a condition or activity, is significant. For Heidegger, temporality is not an entity, not a sequence of self-contained moments that move from future to present to past, and not a property or feature of something, but is, rather, akin to a self-generating and self-transcending process. And, since that process underpins the Being of Dasein, it must be the condition for the possibility of its ecstatic quality – the distinctively human capacity to be at once ahead, behind and alongside oneself, to stand outside oneself, to exist (in grasping the Being of other present beings – its inherent worldliness – and in its self-projective thrownness). In other words, if Dasein's unity as an existing being is literally 'ecstatic' (a matter of Dasein's Being-outside-itself, hence being internally related to what it is not, being non-self-identical), then temporality must be thought of in similarly ecstatic terms. On such a model, past, present and future are not coordinates or dimensions but 'ecstases' – modes of temporality's self-constituting self-transcendence: 'temporality's essence is a process of temporalizing in the unity of the ecstases' (BT, 65: 377).

These claims are only provisional pointers to the full meaning of Heidegger's notion of temporality, which will emerge in later chapters, but they make it clear that this notion bears little relation to common sense or orthodox philosophical conceptions of time. Even if we take it seriously, then, accepting it will violently disrupt our everyday understanding; but such disruption is hardly surprising. After all, the ready glosses or interpretations of time with which our ordinary experience and the philosophical tradition supplies us are all too likely to be the products of inauthenticity – further symptoms of Dasein's flight from an understanding of its own nature rather than useful insights into it. Uncovering an authentic understanding of time and its significance for human life positively requires a violation of such average everyday interpretations.

Nevertheless, no authentic understanding can entirely leave behind its inauthentic rivals. Since they have been embodied in a long history of human thought and human modes of life, they cannot be entirely ungrounded in the ontological realities of Dasein's Being. And, since Dasein cannot entirely lose touch with the meaning of its own Being without ceasing to be Dasein, even its inauthentic conceptions of phenomena cannot be wholly erroneous. A truly ontological investigation of time must therefore show how such inauthentic conceptions – and lives lived out in accordance with them – can emerge from a being to whose Being an understanding of its own nature necessarily belongs. It must show how temporality can temporalize itself inauthentically as well as authentically. The final three chapters of *Being and Time* are devoted to just this task.

First, however, Heidegger must show that his new conception of the internal relation between care and temporality is consistent with, and capable of deepening the insights contained in, his earlier analysis of the various elements that make up the care-structure. He must, in fact, demonstrate that those elements can only be properly understood if they are seen as founded in the tripartite unity of the temporal ecstases – even if the peculiarly ecstatic, self-negating mode of that unity will also put in question any lingering, overly simple conception of Dasein's care-structure as self-identical. At the same time, given that Dasein's existence takes either authentic or inauthentic forms, he also aims to show that both are founded in temporality – indeed, that authentic modes of existence are most fundamentally to be distinguished from inauthentic ones according to the precise mode of temporalizing they manifest. He thus goes over ground that he covered in much detail in the second half of Division One of *Being and Time*, to achieve an even more basic level of understanding – one that changes no specific element, but at the same time radically recontextualizes the entirety, of that earlier analysis.

In following these revisions, we must therefore bear in mind the very different nature of his two aims. For, although both are onto-logically oriented (the first dealing with the existential grounding of such constitutive elements of Being-in-the-world as understand-ing and state-of-mind, the second with the existential grounding of Dasein's capacity to take its own Being as an issue for it), the latter's focus upon the distinguishing temporal marks of authentic as opposed to inauthentic modes of existence naturally requires the use of specific examples of the two modes, and so involves ontic or existentiell analysis. We must be careful not to conflate these two analytical dimensions: we must not confuse the ontic with the ontological, the existentiell illustration with the existential insight.

The elements of the care-structure with which Heidegger concerns himself are: understanding, state-of-mind, falling and discourse. Each is treated separately, but, since they comprise an articulated totality, their internal relations are strongly emphasized and guide the discussion as a whole:

> Every understanding has its mood. Every state-of-mind is one in which one understands. The understanding which one has in such a state of mind has the character of falling. The understanding which has its mood attuned in falling Articulates itself with relation to its intelligibility in discourse.
>
> (BT, 68: 385)

It isn't difficult to see the most obvious sense in which these related aspects of the human way of being have particular facets of temporality as their condition of possibility. The projective nature of the understanding – Dasein's capacity to actualize its existentiell possibilities – is itself possible only for a being that is open to the future. This corresponds to the Being ahead-of-itself of care. Dasein's finding itself always already thrown into moods shows how its present existence is determined by and as what it has previously been, and so presupposes its openness to the past. This corresponds to the already-having-been of care. And the idle talk, curiosity and ambiguity of Dasein's fallenness, understood as modes of its relations with the beings in its environment, could only be attributed to a being that is open to that present environment, and so to the present as such. This corresponds to the Being-alongside of care. Discourse completes the picture, as the articulation of the structures of intelligibility in terms of which the world of this thrown, falling, projective being is disclosed. It thus lacks any links with one particular temporal ecstasis. But the tensed nature of the languages in which discourse has its worldly existence (and which forms so fundamental an aspect of grammatical structures), as well as their capacity to embody truthful claims about the world, would not themselves be possible if the Being of the being who deploys these languages were not rooted in the openness of the temporal ecstasis.

However, even though most elements of the care-structure are primarily associated with a particular temporal ecstasis, properly elucidating the role of that ecstasis will inevitably bring in the other two, and thus an internal relation between any given ecstasis and those which it is not. For example, Dasein's capacity to project itself upon a particular existentiell possibility requires that it utilize the resources of its present environment to do so, and its attunement to the opportunities and constraints that this environment presents is a product of the mood in which it finds itself thrown. Elucidations of moods and falling would take precisely parallel forms; consequently, Heidegger constantly stresses the unity of his conception of temporality, and so the unity of his conception of thrown, projective Being-in-the-world:

> Temporalizing does not signify that ecstases come in a 'succession'. The future is *not* later than having-been, and having-been is *not* earlier than the Present. Temporality temporalizes itself as a future which makes present in a process of having been.
>
> (BT, 68: 401)

Similarly, the vocabulary of 'presuppositions' and 'preconditions' does not mean that temporality provides a kind of framework or medium in which Dasein pursues its existence. Heidegger's idea is not, for example, that Dasein's projections of itself must necessarily be projections into some region or field that we call 'the future'. Rather, just as Dasein's existence *is* projective (projection is not so much something it does as something it is), so its existence is futural (openness to the future is not one of its properties, it is what it is). We are not listing the essential features of a present-at-hand entity, but characterizing a creature who lives a life – a being whose essence *is* existing.

These ideas prepare the ground for Heidegger's second task – that of distinguishing authenticity from inauthenticity in terms of the modes of temporalizing distinctive to each. Once again, he develops his view with respect to each element of the care-structure in turn, and thus focuses on distinguishing authentic from inauthentic modes of the temporal ecstases with which each is primarily associated. But, since the three ecstases are internally related, Heidegger's remarks on each element of the care-structure inevitably contain a portrait in miniature of that which distinguishes authentic from inauthentic modes of temporality in general (in their threefold unity).

Thus, in his examination of understanding, Heidegger defines authentic temporalizing of the future as 'anticipation', and its inauthentic counterpart as 'awaiting'. The former draws on his earlier analysis of anticipatory resoluteness, and amounts to Dasein's letting itself come towards itself out of the future as its ownmost potentiality-for-Being – projecting itself upon whichever possibility best releases its capacity for genuine individuality. By contrast, someone who awaits the future simply projects herself upon whichever possibility 'yields or denies the object of [her] concern'

(BT, 68: 386); the future is disclosed as a horizon from which possibilities emerge that are grasped primarily as either helping or hindering one's capacity to continue doing whatever one is doing in the essentially impersonal manner prescribed by the 'they'.

Both anticipation and awaiting, however, presuppose modes of temporalizing the present and the past. To anticipate the future, Dasein must wrench itself away from its distraction by the present objects of its concern (and, in particular, away from an understanding of its own Being in terms of the Being of such entities), and resolutely determine the present moment as the locus of a concrete existentiell choice. Heidegger talks of this as experiencing a 'moment of vision', in which the resources of the present situation are laid before Dasein in their individual reality and in relation to its own possible individuality. But no such visionary moment is possible without an authentic relation to Dasein's thrownness – without recognizing that one ineliminable aspect of the present situation is the present state of Dasein, and in particular its present attunement to that situation. There can be no authentic appropriation of the future without an authentic appropriation of the past as determinative of the present, and determinative in specific ways. Dasein must acknowledge the past as something not under its control but nonetheless constitutive of who it is, and so as something it must acknowledge if it is to become – to genuinely exist as – who it is. Heidegger labels this 'repetition'; and thus defines authentic temporalizing as an anticipating repetition that holds fast to a moment of vision.

By contrast, the inauthentic mode of awaiting the future presupposes a mode of making present in which Dasein remains absorbed by and dispersed in its environment, disclosing its world in a way dictated by the 'they', which thereby dictates an inauthentic mode of projection. In so doing, it forgets its past – not in the sense that it lacks any awareness of, or overlooks, what has happened to it, but in the sense that it flees from any awareness that what has happened to it is part of who it is. Dasein represses the fact that the existential trajectory which is its life is in large measure determined by the momentum of its particular thrown attunement to the world. It also represses the fact of this repression – the fact that its present dispersal in the 'they' results from its own flight from acknowledging the true basis of its potential for individuality. In this way, inauthentic temporalizing appears as the awaiting which forgets and makes present.

Heidegger's discussion of the other elements of the care-structure attempts to flesh out these general characterizations. In the case of states-of-mind, for example, he contrasts fear and anxiety as illustrative of inauthentic and authentic modes of temporalizing respectively. It might seem that fear is essentially future-oriented and so is a counter-example to the claim that moods primarily presuppose openness to the past; after all, fear of a rabid dog is surely a fear of the threatening possibility that the dog will infect us. The relatedness of any one ecstasis to the other two, however, allows plenty of room for acknowledging that moods must involve a particular relation to the future; but, since moods embody an attunement – a mode of Dasein's openness to its world – they also, and more fundamentally, involve a relation to the past. For example, fear implicates a human being in a mode of forgetfulness. When someone relates fearfully to the future, what she fears for is, of course, herself; and, when she allows such fearfulness to dominate her, the desire for self-preservation dominates her life. She leaps from one possible course of action to another without concretely relating to any of them, her grasp of her present environment dissolves (at best, resolving itself into a bare understanding of entities as handy or unhandy for evading the threat), and she pays no heed whatever to her past. Indeed, the very notion that she has a past, that who she is is determined by who she was and the world in which she found herself, drops away as entirely superfluous in relation to her present goal, which amounts to subordinating everything to the task of continuing to exist and thus to abdicating entirely from the task of determining precisely how that existence might be conducted. She thereby represses the fact that she is delivered over to her own Being as something that is an issue for her – or rather she reduces that aspect of her thrownness to its most nearly animal form. In effect, she allows the possibility of a threat to her life to shatter it entirely. For Heidegger, this is the epitome of inauthenticity, the polar opposite of what is required to live in anticipation of the possibility of one's death, an extreme form of the awaiting which forgets and makes present.

Anxiety, by contrast, makes possible an authentic grasp of one's existence as Being-in-the-world. It is that mood in which Dasein is anxious about its existence in the world in the face of its own worldly existence. Dasein confronts, not a concrete threat to its well-being, but nothing in particular; and this objectlessness confers a merciless perception of the 'nothingness' of the world, of the uncanniness at its core and so at the core of Dasein. When Dasein finds itself in a world whose entities have at present lost any involvement or significance for it, two things are revealed. First, that no given array of entities and circumstances in a given mode of life in itself exhausts the possible significance of Dasein's existence. And, second, that Dasein is nonetheless always already in a world and so forced to choose one existentiell possibility from the array that the world offers. Once again, then, a mood illuminates the essentially enigmatic thereness of Dasein's existence, its existence as thrown and so as open to the past. But, in revealing the actual insignificance of any given world, and so the impossibility of Dasein's ever fulfilling itself by clinging to the present arrangements of its world, anxiety also lights up the world itself as a realm of possible significance, and so the possibility of Dasein's projecting itself upon an authentic mode of existence. In other words, anxiety confronts Dasein with the possibility of its thrownness as something capable of being repeated; and any such repetition is the hallmark of authentic temporalizing.

It is vital to recall here the distinction drawn earlier between existentiell illustrations and the existential insights they illuminate. This analysis of moods does not entail that fearfulness is always inauthentic and anxiety authentic. Although Heidegger does say at one point that 'He who is resolute knows no fear' (BT, 68: 395), it would plainly be absurd (and contrary to the whole thrust of his earlier analysis of moods as genuinely and importantly revelatory of the world) to claim that the authentic man never meets situations in which fear would be the only intelligible response. To fail to take avoiding action when faced with a rabid dog, for example, would be a sign, not of resolution, but of insanity. The point is, rather, that one type of fearful response to genuinely threatening situations is to allow oneself to be entirely overwhelmed by it – to respond like a headless chicken, letting one's attunement to one's world as threatening entirely annihilate one's capacity to grasp its presently definitive lineaments and project the necessary action from among the options available. In so far as fear induces such self-repression or self-forgetfulness it is inauthentic; but not all states of fearfulness fit this description. Similarly, Heidegger never claims that being in a state of anxiety is a criterion for living authentically. On the contrary, he stresses that an anxious grasp of the nothingness at the heart of the world is not in itself a moment of vision. 'Anxiety merely brings one into the mood for a possible resolution. The Present of anxiety holds the moment of vision at the ready; as such a moment, it itself, and only itself, is possible' (BT, 68: 394). It is only if a human being responds to anxiety by actually opening herself to a moment of vision and thereby to anticipating the future by repeating herself from out of the past that authenticity is attained.

Whether this same distinction can be applied to the third main element of the care-structure – falling – is a moot point. Heidegger concentrates upon the mode of temporalizing that underlies curiosity, which he earlier defined as distinctive of falling. This turns out to be an inauthentic temporalizing of the present. To be driven by curiosity is to leap continually from phenomenon to phenomenon, no sooner alighting upon something before definitively consigning it to the past as outmoded and replacing it with something else that attracts one's present concern, only because it is new rather than because of any aspect of its true nature. This is a paradigm case of the awaiting that forgets and makes present, and so a paradigm of inauthentic existence. If, however, falling so defined were an essential element of the care-structure on the same level as understanding and states-of-mind, that would seem to amount to claiming that Dasein was inherently inauthentic – that no mode of its existence could be truly free of lostness in the 'they'. We must, therefore, recall the interpretation argued for earlier, when we examined Heidegger's original treatment of falling. Human existence as worldly thrown projection, and in particular the fact that human beings are primarily located in that world through their occupation of impersonally defined roles, means that lostness in the 'they' is the inevitable default position for Dasein. It can emerge from its lostness by relating to its roles in ways that manifest its individuality; but, in order to do so, it must resolutely wrench itself away from curiosity. In other words, the point of Heidegger's specification of falling as an element of the care-structure is to stress that there is nothing purely contingent or accidental about the prevalence of curiosity, idle talk and ambiguity in Dasein's everyday life; it is *not* intended to suggest that immersion in these existentiell phenomena is somehow necessary or irredeemable. Nevertheless, no one ever finds themselves to have been always already authentic. Authenticity is an achievement:

> Dasein gets dragged along in thrownness; that is to say, as something which has been thrown into the world, it loses itself in the 'world' in its factical submission to that with which it is to concern itself. The Present, which makes up the existential meaning of 'getting taken along', never arrives at any other ecstatical horizon *of its own accord*, unless it gets brought back from its lostness by a resolution.
> (BT, 68: 400; my italics)

THE TEMPORALITY OF CARE: BEING IN THE WORLD (§§69–70)

With this account of the temporal basis of falling, Heidegger's doubly motivated analysis of the various elements of the care-structure is in one sense complete. But each element has only a relatively autonomous life, so he ends by stressing the priority of the articulated unity of the care-structure. This returns us to an even earlier stretch of his analysis of the human way of being. For the first division of *Being and Time* showed that the care-structure grounds Dasein's existence as Being-in-the-world – its always already being in a world in which it can encounter entities as the kind of entities they are. So, if the basis of the care-structure as a whole is temporality, Dasein's openness to beings in the world – its capacity to reach beyond itself to that which is not itself – must itself have an essentially temporal grounding. In short, Dasein's existence as ecstatic Being-in-the-world must be based upon the threefold ecstasis of temporality.

Heidegger's earlier analysis of Dasein's everydayness focused upon its relations with objects as handy or unhandy for its practical activities. It also stressed that encountering any object as a piece of equipment presupposed an equipmental totality, i.e. that no individual tool could be encountered as such except against the background of an array of other items – a pen exists as a pen only in relation to ink, paper, table and so on. Such arrays are themselves grounded in a set of assignment-relations: the utility of a tool presupposes something for which it is usable (its 'towards-which'), something from which it is constructed and upon which it is employed (that 'whereof' it is made), and a recipient for its end product. This web of socially constituted assignments – 'the world' – founds the readiness-to-hand of an object; but it is itself founded in a reference to particular projects of Dasein's – the handiness of a hammer, for example, being ultimately a matter of its involvement in building a shelter for Dasein. In short, the ontological basis of the world (its worldhood) lies in specific possibilities of Dasein's Being. But Dasein's relations with specific existentiell possibilities presuppose its existence as thrown projection – possessed of understanding, possessed by moods; and these elements of the care-structure have temporality as their condition of possibility. It follows that the basis of Dasein's openness to entities is its openness to past, present and future: for Dasein to disclose entities is for it to manifest a present concern for them, which grows from its having taken on a project and being oriented towards its future realization. Dasein's worldliness is thus grounded upon the temporalizing of temporality.

Of course, Heidegger's earlier account focused upon Dasein's average everyday modes of encountering objects as ready-to-hand, and so upon an inauthentic mode of its existence – one in which Dasein has succumbed to its inherent tendency to lostness, to a fascination with the objects of its concern which elides its non-identity with them. So the specific mode of temporalizing presupposed in average everydayness is fundamentally inauthentic. Average everyday Dasein relates to its work by forgetting itself, entirely subordinating its individuality to the impersonal requirements of its task. So it represses its pastness rather than repeating or recovering it, its concern for the objects in its environment makes them present in entirely irresolute ways rather than facilitating a moment of vision, and the goal of its labours is determined by the anonymous expectations of the public work-world rather than by its responsibility to become a genuine individual. In short, average everyday Being-in-the-world is a making-present which awaits and forgets; but not all Being-in-the-world – and, in particular, not every interaction with objects as ready-to-hand – is so grounded.

The temporal basis of Dasein's Being-in-the-world is equally evident when Dasein holds back from practical engagements with entities and encounters them instead as present-at-hand – for example, in the context of scientific study. For the objects concerned are not then encountered outside or independently of the world and its ontological structures. True, in such a transformation of Dasein's relations with objects, the specific work-world and the specific existentiell project that provided the original context for its concern with them disappears; a hammer originally encountered as a tool for building a house is then confronted as a material object possessed of certain primary and secondary qualities. But this is not a matter of de-contextualizing the object, but of re-contextualizing it; the scientist embeds it in a very different web of assignment-relations, but it remains no less embedded in a world for all that. As we suggested in Chapter 1, and as Heidegger now emphasizes:

> Just as *praxis* has its own specific kind of sight ('theory'), theoretical research is not without a praxis of its own. Reading off the measurements which result from an experiment often requires a complicated 'technical' set-up for the experimental design. Observation with a microscope is dependent upon the production of 'preparations' . . . even in the most 'abstract' way of working out problems and establishing what has been obtained, one manipulates equipment for writing, for example. However 'uninteresting' and 'obvious' such components of scientific research may be, they are by no means a matter of indifference ontologically.
>
> (BT, 69: 409)

In other words, scientific investigation is not a purely intellectual matter: it does not require the complete suspension of praxis. Rather, it substitutes one mode of praxis – one mode of concern for objects, one mode of letting them be involved in Dasein's projects – for another. Encountering objects as present-at-hand is a particular mode of Being-in-the-world. The disclosure of entities as physical objects does not reveal that which makes possible the existence of Dasein in a world (by revealing the essential nature of that world), but is itself only possible because Dasein's existence is worldly (and thus capable of disclosing entities at all). Science, too, involves making objects present in a particular kind of way (thematizing them as present-at-hand) in the context of a specific human enterprise (that of grasping the truth about beings understood as physical phenomena), and so in relation to a particular possibility of Dasein's Being (namely, its Being-in-the-[scientific-]truth). It therefore presupposes the seeing-as structure of disclosedness, which is itself grounded in some mode or other of temporalizing. 'Like understanding and interpretation in general, the "as" is grounded in the ecstatico-horizonal unity of temporality' (BT, 69: 411).

There is thus more to the human way of being than is manifest in any particular encounter with or thematization of specific entities – it is Being-in-the-world. And Heidegger's final question in this chapter is: what must be the case for this ontological truth about Dasein to be possible? What kind of existence or Being must the world have if Dasein's Being is inherently worldly? What is the true nature of the link between Dasein and the world? The short version of his answer is this: Dasein exists as Being-in-the-world because the Being of Dasein is transcendence and so is that of the world, and the basis of that transcendence in both cases is temporality. The longer answer goes as follows.

As thrown, falling projection, Dasein is transcendent in the sense that it is always more or other than its actual circumstances and form of life: it relates itself to possibility rather than actuality – its present state is the basis for projecting upon an existentiell possibility once it has appropriated the past as determinative of what it now is. The world is transcendent in the sense that it is something more or other than the Being of any actual entities within it. It is not an entity but a web of assignment-relations within which any specific object is encounterable as ready-to-hand or as present-at-hand,

and without reference to which neither readiness-to-hand nor presence-at-hand as such could be understood. The basis of Dasein's transcendence is temporality: thrown projection is the mode of existence of a being open to past, present and future. The basis of the world's transcendence is also temporality: since the world constitutes an arena for disclosing objects in terms of (i.e. assigning them to) a particular mode of practical activity, it must be capable of accommodating the essentially temporal references of any praxis – in which objects are presently taken up in the course of an already initiated task and in a manner determined by its projected completion. In other words, the world as entity-transcendent exists as the field or horizon within which Dasein realizes itself as a self-transcending actualizer of possibilities. And what underwrites the complementarity of Dasein's horizon-presupposing transcendence and the world's horizon-providing transcendence is the ecstatic (i.e. horizonal) threefold unity of temporality.

Thus, the temporal ecstases play a role in Heidegger's analysis that parallels Kant's invocation of schematism in the Transcendental Deduction of his *Critique of Pure Reason*.[1] Having defined the categories (pure concepts of the understanding) in terms of logical principles, and having argued that no experience of objects is possible unless the manifold of intuition is synthesized by means of those categories, Kant needs to show how such pure concepts might conceivably be commensurable with what seems entirely heterogeneous to them, namely the chaotic matter delivered up by the senses. He engineers this transition from pure categories to categories-in-use by positing the existence of a set of schemata, each of which is what he calls a 'monogram of pure a priori imagination' – a pure synthetic rule couched in terms of temporal ordering (the most general form of sensible intuition, on Kant's account). Each such schema, in so far as it is a rule, has a recognizable kinship with a purely logical relation; and, in so far as it is a rule of temporal order, it also has application to sensibility. Schemata are therefore essentially Janus-faced – at once possessed of the purity of the a priori and the materiality of intuition: as the nexus of concepts and intuitions, they form the junction-box through which the Kantian system relates mind and matter, subject and world.

Heidegger registers these Kantian echoes by claiming that to each of his three temporal ecstases there belongs a 'horizonal schema' – a 'whither' to which Dasein is carried away or dragged out. With the future, it is 'for-the-sake-of-itself'; with the past, it is 'what-has-been'; with the present, it is 'in-order-to'. These glosses recall elements of the structure of significance that constitutes the world-hood of the world upon which Dasein projects itself, and so confirm that Heidegger's schemata are a response to precisely the difficulty facing Kant – that of demonstrating the essential complementarity of human subject and objective world. To this degree, Heidegger acknowledges that Kant preceded him in identifying a significant ontological problematic and in at least pointing towards the key concept needed to address it. But he does not take himself to be addressing the problem in exactly the way Kant does.

To begin with, in so far as Kant's account rests upon his analysis of time as a form of sensible intuition, it draws upon his more general assumption of a distinction between the form and the content of experience; its content is elucidated in terms of present-at-hand representations, and its form as something imposed by the synthetic activities of the transcendental subject. Heidegger explicitly rejects the terms of this account:

> The significance-relationships which determine the structure of the world are not a network of forms which a worldless subject has laid over some kind of material. What is rather the case is that factical Dasein, understanding itself and its world in the factical unity of the 'there', comes back from these horizons to the entities encountered within them.

> (BT, 69: 417)

For Heidegger, the Kantian account of experience entirely fails to distinguish between entities and the world within which they are encountered, and so loses any chance of coming to understand Dasein as Being-in-the-world. Heidegger's temporal schemata are not entities or structures that mediate between the otherwise inde-pendent elements of Dasein and world. For him, human Being and world are primordially and indissolubly united, and his account of temporality as its basis is rather an attempt to locate the single root from which the twofold articulation of Being-in-the-world must grow, if that hyphenation truly registers a differentiation within a fundamental unity rather than a conjunction.

Moreover, the ground (and so the nature) of that fundamental unity must be understood in ecstatic rather than static terms. Where Kant compares his schemata to monograms, Heidegger talks of his as horizons whither Dasein is always already carried away or dragged out, since it could not otherwise come back to confront entities that necessarily appear within those horizons. Each horizonal schema thereby indicates an aspect of Dasein's worldly Being as standing-outside-itself, one respect in which Dasein's distinctive mode of identity (and hence that of its world) is one of non-self-coincidence. Accordingly, one must understand the fundamental unity of Dasein and world with which Kant was so concerned – their inherent aptness for one another – as a function of their individual non-self-identity; the internal relation between Dasein and world is generated by the internal self-differentiation of Dasein and of its world. One might say: Dasein's failure to coincide with itself and its openness to what it is not are ultimately indications of one and the same phenomenon – its temporality.

These connections and contrasts with Kant's investigation are sufficiently important for Heidegger to conclude his analysis of everydayness and temporality by developing a further analogy – one involving Dasein's spatiality. The fundamentality of time in his account of Being-in-the-world might suggest that Heidegger has overlooked or insufficiently appreciated the deep importance of the notion of space to our conception of the world. But Heidegger's view is that, although Dasein's spatiality is indeed fundamental, it is nonetheless subordinate to its temporality.

The Kantian echo here is of the priority Kant famously assigns to time over space. Kant defines both as forms of sensible intuition – not elements within that manifold but rather the two modes through which those elements are always and necessarily experienced by us as interrelated. But, while our experience of the external world is both spatially and temporally ordered, our experience of our inner world, of the ebb and flow of our thoughts, emotions and desires, is ordered only temporally. Since our representations of the external world are themselves necessarily a part of our inner world (consequences of our being affected by the senses), time, as the form of inner (and therefore of outer) sense, trumps space, which is merely the form of outer sense.

Once again, Heidegger implicitly acknowledges the grain of truth in Kant's analysis by vehemently condemning the details of its working out:

> If Dasein's spatiality is 'embraced' by temporality . . . then this connection . . . is also different from the priority of time over space in Kant's sense. To say that our empirical representations of what is present-at-hand 'in space' run their course 'in time' as psychical occurrences, so that the 'physical' occurs mediately 'in time' also, is not to give an existential-ontological Interpretation of space as a form of intuition, but rather to establish ontically that what is psychically present-at-hand runs its course 'in time'.
>
> (BT, 70: 419)

Unlike Kant, who fails to attain a genuinely ontological level of analysis because he assumes that our experience of objects consists of present-at-hand representations of them, Heidegger sees that Dasein's spatiality is existentially founded upon its temporality. Although practical activity in the world presupposes spatiality, the modes of spatiality thereby disclosed can only be elucidated by reference to the temporal foundations of the worldhood of the world:

> Whenever one comes across equipment, handles it, or moves it around or out of the way, some region has already been discovered. Concernful Being-in-the-world is directional – self-directive. . . . [But] relationships of involvement are intelligible only within the horizon of a world that has been disclosed. Their horizonal character, moreover, is what first makes possible the specific horizon of the 'whither' of belonging somewhere regionally . . . a bringing-close (de-severing) of the ready-to-hand and the present-at-hand [is] grounded in a making-present of the unity of that temporality in which direction-ality too becomes possible.
>
> (BT, 70: 420)

Dasein's spatial existence is primarily a matter of placing itself in relations of proximity to, and distance from, objects according to the demands of its practical activities; so it presupposes the disclosure of a work-world and so of the world as such, which is founded in the horizonal ecstases of temporality.

REPETITION AND PROJECTION (§71)

Heidegger concludes his chapter by declaring that he has not yet fully penetrated the existential-temporal constitution of Dasein's everydayness – a deflating declaration for any reader who has struggled with what seemed to be exhaustive (and exhausting) revisions of the provisional insights into everydayness expressed in Division One. But it is undeniable that the very term 'everydayness' has temporal connotations which are as yet unexplored. It variously suggests an idea of human existence as a sequence of days, of the daily or the diurnal progress of time, of its being marked by habitual, customary or repetitive experiences, attitudes and practices that both maintain themselves and alter across the wider stretches of time that make up the periods of human history. In other words, Dasein's relation to temporality necessarily involves it in the daily round of everyday life and in the passage of time more broadly understood, in history; and these are the topics of Heidegger's final two chapters.

The present chapter thereby acquires a very distinctive pattern, one which emerges when we step back from its details and view it as an articulated whole. The chapter begins from a sense that our general investigation of the Being of Dasein has reached a pivotal point – a moment of insight into the temporal grounding of the care-structure, and so to a view of the various elements of human conditionedness or finitude as themselves conditioned by temporality. It presents that insight as requiring a return to the material outlined earlier in *Being and Time*, a return that the chapter itself enacts in order to show that this insight at once deepens, unifies and radically recontextualizes our understanding of the claim that Dasein is Being-in-the-world. And it ends by outlining the ways in which this repetition of past claims delivers a fruitful direction for further investigation.

The emphasis upon retracing one's steps that this chapter structure enacts is exactly what one should expect from a philosopher who has made much of the essentially circular nature of understanding and interpretation. For, if all human comprehension is always already inside a hermeneutic circle, motivated by some particular structure of fore-having, fore-sight and fore-conceptions, then one can only make progress in one's philosophical understanding by retracing one's steps within the circle and deepening or modifying one's grasp of the elements of one's fore-structure. But then the second time around the circle (being temporally distinct from its predecessor) is in fact the second turn of a spiral, and hence should not be thought of as a simple retracing of one's steps. After all, such retracings are always the act of a being whose Being is Being-guilty, hence the null basis of a nullity, so no Dasein could ever completely sweep up its earlier, past steps into its own present comprehension. And it is precisely this lack of absolute coincidence between past and present that opens up the possibility of grasping new reaches of significance; absolutely exact recapitulations of past understandings would make progress in human understanding inconceivable.

Hence, Heidegger's restatements of his earlier, provisional conclusions can never exactly coincide with them; he could never succeed in simply saying again, even if at a deeper level, exactly and only what they said, but will rather say them otherwise, placing them in a new context of considerations – above all, in the context provided by a realization of the general significance of this phenomenon of non-self-coincidence (and hence of Dasein's internal relation to nothingness) for any proper grasp of Dasein's Being. Hence, the uncanny sense that Heidegger's revisioning of his earlier vision of the human way of being at once confirms and subverts that vision; for it shows us that his earlier vision missed nothing in particular, and yet that everything in the initial vision seems utterly different when grasped in its inherently enigmatic relation to that nothing.

However, the structure of this chapter is more distinctive than hermeneutic circularity or spiralling would require; or, at least, its distinctiveness is overdetermined. For, if one had to summarize that structure in a single sentence, a structure through which Heidegger's key insight into the grounding role of temporality generates a rewriting of his earlier discoveries with a view to moving his project forward, one might say that it is an anticipating repetition which holds fast to a moment of vision. In other words, the experience of reading it has an underlying ecstatic temporal structure that precisely fits Heidegger's definition of authentic temporality. The composition of the chapter enacts the structure of its topic: the movement of Heidegger's prose declares its own authenticity as a piece of writing and attempts to elicit an act of authentic reading from those it addresses. Once again, the form and the content of *Being and Time* are mutually responsive: the understanding of human existence to which its propositions lay claim determines a conception of the proper relation between author and reader that is reflected and enacted in its form.

NOTE

1   Kant, *Critique of Pure Reason*, trans. N. Kemp Smith (London: Macmillan, 1929).

7

FATE AND DESTINY: HUMAN NATALITY AND A BRIEF HISTORY OF TIME

(Being and Time, §§72–82)

HISTORY AND HISTORICALITY (§§72–5)

Heidegger claims that everyday human existence is diurnal – lived out daily, from day to day, every day; Dasein is stretched along in the sequence of its days. The notion of Dasein being *stretched along* is implicit in the care-structure and the temporality-structure that underlies it. Since Dasein exists as thrown and projecting (not as something initially self-identical that is then stretched out but rather a being that is always already ahead of itself and always already having been), Heidegger's earlier claim that Dasein exists as 'the Being of the between' must have a temporal connotation. The human openness to the world depends upon an openness to time – upon the fact that human beings exist as temporality, that the human way of being is ecstatic temporalizing. Now, however, Heidegger reformulates this claim:

> The specific movement in which Dasein is *stretched along and stretches itself along*, we call its '*historizing*'. . . . To lay bare the *structure of historizing*, and the existential-temporal conditions of its possibility, signifies that one has achieved an ontological understanding of *historicality*.
>
> (BT, 72: 427)

Why this shift from talk of temporalizing and temporality to talk of historizing and historicality? Heidegger's account of Dasein's temporality has thus far accorded a certain priority to its existence as futural, to 'Being-ahead-of-itself'; in outlining the structure of anticipatory resoluteness, and so of authentic human existence, he placed the human capacity to project, to relate oneself to one's own end, at centre stage. If everydayness is a stretching along between birth and death, an emphasis on death has tended to eclipse birth. But, if Dasein really is the Being of this between, then it is just as fundamental to its Being that it exists as born as that it exists as always already dying. If no temporal ecstasis can be separated from the other two, then Dasein's pastness must inflect its relation to present and future, and so inflect its temporalizing more generally. But, then, what it is for Dasein to exist as a historical being, what it might mean to say that Dasein has a past or can relate to the past, or to say that in so far as Dasein exists it historizes, must be elucidated in the terms of our earlier analysis of temporality. For only a creature whose way of being is essentially temporal could live a life that is essentially historical in these several ways.

Particular historical findings will cast no light on the question of Dasein's historicality – for any results of historical investigation will presuppose precisely what is at issue here, namely the human ability to explore the past. Furthermore, on Heidegger's view, no previous study of history as a science or discipline (no historiology) has properly engaged with its subject matter because none has taken a fully existential-ontological perspective on this activity of Dasein. None has asked about the conditions for the possibility of history *and understood that discipline as one activity of a being whose way of being is inherently worldly*. Accordingly, he intends to elucidate the temporal significance of Dasein's existence as *thrown* projection by probing the significance of its existence as historical.

This means breaking up the average everyday understanding of Dasein's historicality, and of historicality more generally. When inauthentically oriented, human beings interpret the question of their own historicality as a matter of explaining the possibility of their own connectedness through time – showing how a single continuous self can persist unscathed through a sequence of temporal moments that appear from the future, become the present and then disappear into the past. This is certainly the form in which this question has been posed in the modern tradition of philosophy from Hume to Parfit.[1] For Heidegger, such interpretations assume that time is a collection of self-contained units that begin by being not yet present-at-hand, become momentarily present-at-hand and then become no longer present-at-hand; and human beings are seen as dispersed in them, scattered across a sequence of past, present and future nows, and in need of unification. Similar atomistic assumptions are at work when the historicality of events and objects is under consideration. A past event is one that has happened and is now irretrievably lost, a historical object something that was once at hand but is so no longer. Even if a given event continues to have significance for our present world, it is understood as a piece of the past that has consequences in the present (in the way that a past cause can have contemporary effects) – just as a historical arte-fact in a museum is thought of as a piece of the past that remains present-at-hand.

Heidegger attacks this picture of historicality at what might seem its strongest point – the claim that the historicality of an object (for example, a household implement in a museum) is a matter of its being something that belongs to the past but is present-at-hand in the present. For, if the historicality of an object is a matter of its belonging to the past, and the past is understood as those moments of time that are no longer present-at-hand to us, how can an object that is still present to us nonetheless be something historical? Such antiquities must somehow embody pastness, must be marked by and so manifest the passage of time. But what is this mark of past-ness? An ancient pot or plate is likely to have altered over time – becoming damaged or perhaps simply more fragile; but such wear and tear cannot be what makes them historical, since contemporary objects suffer the same indignities, and an undamaged object from the past is not thereby rendered contemporary. Nor can their pastness consist in the fact that they are no longer used for the purposes for which they were originally designed; a dinner plate passed down from generation to generation is no less an heirloom simply because it is still used on special occasions to serve food. Nonetheless, such a plate used in such a way is somehow altered, no longer what it was; something about it belongs to the past – but what?

> Nothing else than that *world* within which they belonged to a context of equipment and were encountered as ready-to-hand and used by a concernful Dasein who was-in-the-world. That *world* is no longer. But what was formerly *within-the-world* with respect to that world is still present-at-hand.

> (BT, 73: 432)

The dinner plate belongs to the past because it belongs to a past world. It constitutes a trace of a particular conceptual and cultural framework within which it fitted as one element in a totality of equipment suitable for one type of human activity – one involving the ingestion of sustenance, but also the provision of hospitality, the maintenance of family life, the preservation of a complex of cultural practices, and so on. It remains present to us as an object within our world, and – whether used to serve food or displayed in a cabinet – as a ready-to-hand item within that world (ready-to-hand as a piece of domestic crockery or an antiquity). But it is still an heirloom, still an historical object, because it is marked by the world for which it was originally created and within which it was originally used. Even for the family for which it is an heirloom, it is not used for serving food in just the way their contemporary dinner service is used – the heirloom is for special occasions.

If the worldliness of historical objects is what constitutes their pastness, then that pastness is doubly derivative: the condition for its possibility is the past existence of a world, and the condition for the possibility of such a world is the past existence of Dasein (the being whose Being is essentially worldly). In other words, the historicality of objects and events is derivative of the historicality of

Dasein; Dasein is what is primarily historical. But the pastness of Dasein cannot be understood in terms of presence-at-hand or readiness-to-hand. 'Past' Dasein is not an entity who was, but is no longer, either present-at-hand or ready-to-hand. It is a being who existed but no longer does so, a being who *has been* – a being whose Being is existence. So human beings do not become historical only in so far as they no longer exist; historicality is not a status they achieve only when they die. On the contrary, a being who exists as Being-in-the-world must exist as ecstatic temporalizing, as transcending itself in the threefold unity of the ecstases, and so as open to the past. A worldly being is something futural that has been and is making present, and so is a being that always already *has been*. In short, for Dasein to exist at all is for it to be historical.

Heidegger's exploration of this issue is dominated by the question of Dasein's authenticity. Since Dasein's Being is an issue for it, its modes of existence are either inauthentic or authentic; and, if its existence is inherently historical, there must be inauthentic and authentic modes of its historizing. The authentic mode must embody anticipatory resoluteness – a projecting which is reticent and ready for anxiety. But any projecting presupposes a range of available existentiell possibilities upon which to project; and this raises the question of whence Dasein can draw these possibilities. They cannot be provided by its death, by Dasein's Being-toward-its-end; projecting upon that possibility guarantees only the totality and authenticity of its resoluteness. We must look instead towards the other pole or dimension of Dasein's stretching along – to its birth rather than its death, or, more precisely, to its thrownness.

As thrown, Dasein is delivered over to a particular society and culture at a particular stage in its development, in which certain existentiell possibilities are open to it and certain others not: becoming a Samurai warrior, a witch or a Stoic are not available options for early twenty-first-century Westerners, whereas becoming a police officer, a social worker or a priest are. Dasein is also thrown into its own life at a particular stage in its development, which further constrains the range of available choices. One's particular upbringing, previous decisions and present circumstances may make becoming a social worker impossible or becoming a priest almost unavoidable. In other words, the facts of social, cultural and personal history that make up an individual's present situation constitute an inheritance which she must grasp if she is to project a future for herself; and part of that inheritance is a matrix of possible ways of living, the menu of existentiell possibilities from which she must choose. She can do so inauthentically – understanding herself 'in terms of those possibilities of existence which "circulate" in the "average" public way of interpreting Dasein today [and which] have mostly been made unrecognizable by ambiguity [although] they are well known to us' (BT, 74: 435); or authentically – in which case she resolutely 'discloses current factical possibilities of authentic existing, and discloses them *in terms of the heritage* which that existence, as thrown, *takes over*' (BT, 74: 435).

Defining authentic appropriations of one's thrownness as taking over a heritage carries a field of interlocking connotations. First, the average everydayness from which everyone always begins is itself part of one's heritage: Dasein is always delivered over to lostness in the 'they', and so to the average public way of interpreting the available existentiell options that its social and personal culture bequeaths. The prevailing modes of ambiguity and curiosity make these options unrecognizable – covering over their true contours either by making them the focus of an endless debate fuelled by superficial curiosity, or by taking one superficial interpretation of them for granted. Thus, to inherit them properly means seizing upon that heritage in a manner which discloses its true lineaments; it means reacting against one's heritage in order to uncover it properly, reclaiming it. But Dasein must also relate those options to its own individual circumstances and life; it must reclaim itself as its heritage. Lostness in the 'they' involves a dispersal of oneself amid the currents of ambiguity and curiosity. So resolutely taking over one's heritage means rejecting the possibilities that *seem* closest (where that proximity is a function of their ease or acceptability to others) and grasping those that relate to one's ownmost potentialities – the possibilities that resoluteness reveals to be non-accidentally closest to one in the light of an anticipation of one's death.

The heritage of one's culture and the heritage of oneself thus fuse in a mutually revivifying way. An individual's self-constancy in actualizing certain forms of life at once renews the life of those forms and so of the culture that they constitute, and reveals them as capable of defining genuinely authentic individual lives, as possibilities for which individuals are destined and to which they can relate as fateful for themselves and others:

> Once one has grasped the finitude of one's existence, it snatches one back from the endless multiplicity of possibilities which offer themselves as closest to one – those of comfortableness, shirking and taking things lightly – and brings Dasein in to the simplicity of its *fate*. This is how we designate Dasein's primordial historizing, which lies in authentic resoluteness and in which Dasein *hands* itself *down* to itself, free for death, in a possibility which it has inherited and yet has chosen.
>
> (BT, 74: 435)

This is a vision of the freedom available to a conditioned or finite being – a vision of mortal freedom as essentially finite or conditioned (what Heidegger would call an aspect of Being-guilty). Dasein's capacity to choose how to live and who to be is real and distinctive. But it cannot choose not to have that capacity; it must exercise it in circumstances that it has not freely chosen, upon a range of possibilities that it has not itself defined, and on the basis of an understanding of its situation that is itself situated (hence inherently subject to limitations). So it is a power that is necessarily rooted in powerlessness – a freedom founded in abandonment. Its fulfilment thus comes not through any attempted abolition or transcendence of those constraints, but through a resolute acceptance of them as they really are – through a clear acknowledgement of the necessities and accidents of one's situation as one's fate.

And, since fateful Dasein, as Being-in-the-world, is also Being-with-others, its authentic historizing is also what Heidegger calls a 'co-historizing'. The world it inherits is a common and a communal world; the existentiell possibilities that the world offers are bequeathed to individuals through essentially social structures and practices, and typically can only be taken up by them in concert with others. But, by the same token, those structures will only persist if individuals continue to commit themselves to the possibilities they embody; and the culture they constitute will only persist in a vital and authentic way if individuals grasp those possibilities authentically. In other words, Dasein's historizing is at once an individual and a communal affair. To the individual driven about by accident and circumstance, there corresponds a community persisting as the homogenized aggregation of the 'they'; and to the fate of an individual there corresponds the destiny of a people:

> Our fates have already been guided in advance, in our Being with one another in the same world and in our resoluteness for definite possibilities. Only in communicating and in struggling does the power of destiny become free. Dasein's fateful destiny in and with its 'generation' goes to make up the full authentic historizing of Dasein.
>
> (BT, 74: 436)

The risk of emphasizing the natality rather than the fatality of Dasein is that it will appear essentially backward-looking, and thus conservative – as if taking over one's heritage is a matter of mechanically reiterating forms of life and formations of culture lying in the past of the society concerned, thus condemning both individuals and their culture to a living death. There seems little room for reform, innovation or responsiveness to altered circumstance. But this interpretation forgets that hermeneutic understanding takes a spiralling form, so that no new turn around it coincides with its predecessor; and it assumes that historizing is a substitute or a synonym for temporalizing, rather than one aspect of that process. As such, it is inextricably related to the other two temporal ecstases, and so forms part of an articulated unity that also involves a resolute grasp of the present situation and an anticipatory projection into the future. Consequently, what Heidegger calls 'the struggle of loyally following in the footsteps of that which can be repeated' (BT, 74: 437) does not mean binding the present to what is already outmoded. Any reclaiming of one's heritage must flow from a resolute projection into the future based on a moment of vision with respect to the present. So it is better thought of as a reciprocative rejoinder to a past existentiell possibility – a dialogue between past and present, a creative reworking of that possibility in the light of an essentially critical disavowal of the superficialities and ambiguities of what passes for the working out of the past in average everyday life.

Nevertheless, the entanglement of historizing with projection does not entail a simple endorsement of progress; authentic Dasein is as indifferent to novelty as it is to nostalgia. Authentic projection into the future presupposes the taking over of one's heritage, and so is essentially constrained and guided by that inheritance. But the ultimate purpose of reclaiming the past is to project it into the future; and this involves a mode of repetition that acknowledges both the necessities of the present and the genuine potential of the future. Such repetition is an essential component of anticipatory resoluteness, the authentic mode of human temporalizing. We can therefore say, with Heidegger, that '*Authentic Being-towards-death – that is to say, the finitude of temporality – is the hidden basis of Dasein's [authentic] historicality*' (BT, 74: 438). Or, rather more elaborately, but in a way that manifests the underlying unity of the whole of Heidegger's analysis of temporality in Division Two of *Being and Time*:

> Only an entity which, in its Being, is essentially *futural* so that it is free for its death and can let itself be thrown upon its factical 'there' by shattering itself against death – that is to say, only an entity which, as futural, is equiprimordially in the process of *having-been*, can, by handing down to itself the possibility it has inherited, take over its own thrownness and be in the *moment of vision* for 'its time'. Only authentic temporality which is at the same time finite, makes possible something like fate – that is to say, authentic historicality.
>
> (BT, 74: 437)

So much for authentic historizing. The typical mode of Dasein's everyday existence, however, is inauthentic – and such lostness in the 'they' is no less historical. When human beings are lost in the 'they', their historicality and the historicality of their world is not annihilated but repressed – and in two stages. First, Dasein understands its own historicality in terms of the historicality of that with which it is absorbed in its world (i.e. it understands itself world-historically rather than understanding world-historicality as a function of its own historicality); and, second, it interprets that world-historicality in terms of presence-at-hand. Inauthentic Dasein understands the historicality of objects as the appearance and disappearance of present-at-hand entities, and then interprets its own existence according to that model – as a sequence of moments that become present-at-hand and then slip away into the past.

Accordingly, when the question of Dasein's historicality gets raised in philosophy, it is formulated as a matter of determining the connectedness of a series of experiential atoms over time. This is wholly inappropriate to a being whose temporal unity is really a matter of its stretching along and being stretched along between birth and death. But it is an appropriate response to the existentiell situation of a Dasein lost in the 'they' – for such lostness is in one sense a matter of self-inconstancy, of the self being dispersed or dissipated in the shifting currents of ambiguity, curiosity and idle talk. In that sense, a recovery of unity, a pulling oneself together, is required if inauthentic existence is to be transformed into authentic individuality; but any such transformation must be based on an understanding of that unity as the articulated unity of the care-structure, which must itself be grasped in terms of inherently ecstatic temporalizing. Thus, there is more than a grain of truth in the inauthentic conception of the self as requiring connectedness: for whether the individual will take over her fate and the destiny of her people, or instead forget her heritage and the possibilities it opens up, is in reality a question of whether or not she will achieve self-constancy. But self-constancy is not self-identity; and, in particular, it is not a matter of the self's aspiring to, or achieving, identity with its past, but, rather, of its finding openness to a genuine future in its non-coincidence with its past:

> With the inconstancy of the they-self Dasein makes present its 'today'. In awaiting the next new thing, it has already forgotten the old one. The 'they' evades choice. Blind for possibilities, it cannot repeat what has been, but only retains and receives the 'actual' that is left over, the world-historical that has been, the leavings, and the information

about them that is present-at-hand. Lost in the making-present of the 'today', it understands the 'past' in terms of the 'Present'. . . . When . . . one's existence is inauthentically historical, it is loaded down with the legacy of a 'past' which has become unrecognizable, and it seeks the modern. But when historicality is authentic, it understands history as the 'recurrence' of the possible, and knows that a possibility will recur only if existence is open for it fatefully, in a moment of vision, in resolute repetition.

(BT, 75: 443¹)

THE LESSONS OF HISTORY (§§76–7)

Heidegger next shifts the focus of his investigation from historicality to historiology – the science of history. His immediate aim is to demonstrate that it is only because Dasein's existence is historical that it can engage in historical investigation. In one sense, of course, this conclusion follows immediately: if Dasein's existence is historical, then everything it does is grounded in its historizing, and that will be as true of the historian's activities as it is of the carpenter's or the musician's. But, for Heidegger, historiology is more closely and distinctively linked to historicality than this.

If the pastness of phenomena is derivative of the pastness of their world, then an understanding of the past is available only to beings capable of understanding worlds and understanding them as past; and that is possible only for beings whose Being is worldly and open to pastness – that is, for human beings:

Our going back to 'the past' does not first get its start from the acquisition, sifting and securing of [world-historical] material; these activities presuppose *historical Being towards* the Dasein that has-been-there – that is to say, they presuppose the historicality of the historian's existence.

(BT, 76: 446)

In other words, Dasein's capacity to engage with the past is dependent upon its historicality; the very possibility of historiology depends upon the historicality (and so the temporality) of the human way of being.

But the picture Heidegger paints is more complicated than this. For the historicality of objects, events and institutions is itself derivative of the historicality of Dasein. Their pastness depends upon the past existence of a world, which is in turn dependent upon Dasein's having lived in a certain way at a certain time in the past. Thus, the primary object of historical investigations is really Dasein itself – Dasein as past: remains, monuments and records are in effect possible material for the concrete disclosure by existing Dasein of the Dasein which has-been-there. The disclosure of the past is the disclosure of a past world, and thus of a past disclosure of the world; engaging in history is a matter of Being-in-the-world recovering or recreating a past mode of Being-in-the-world, and doing that historical task properly means capturing that past mode of Being-in-the-world as it really was – understanding the past in terms of the real potentialities and limitations of then-prevailing forms of human life.

Accordingly, the true object of historical investigation is not the facts of a past era but a possible mode of existence: true history concerns not actualities but possibilities. But the genuine disclosure of what has-been-there. the recovery of the real potential of a past existentiell possibility, is precisely what Heidegger has been sketching in as the core of authentic human historizing. To understand the Dasein which has-been-there in its authentic possibility just *is* to repeat its mode of worldly existence – to make it available as something handed down to Dasein in its present situation.

This implies that authentic human existence presupposes authentic historiology. For, if Dasein can exist as authentic historizing only by repeating one of its inherited existentiell possibilities, then, whatever mode of life it enacts, it must have recovered its authentic lineaments from the past of its culture. Whether Dasein exists authentically as a historian, a carpenter or a musician, it can do so only by either possessing or drawing upon the skills of the true historian. Since authentic temporalizing involves tearing oneself away from the falling anonymity of the 'they' and its superficial interpretations of available modes of life in the name of a genuinely destined future, its critique of the present must be guided by a disclosure of the true heritage of existentiell possibilities from which an individual and a community can project that future; but such a disclosure is precisely what a properly conducted historical investigation can alone provide.

If, however, authentic historizing presupposes authentic historiology, authentic historiology also presupposes authentic historizing. To realize the true potential of historical investigation, the historian must reveal by repetition the Dasein which has-been-there in its essential possibility. But any such repetition must be guided by correspondingly authentic modes of openness to past and future: to disclose that past possibility as it really was is to reveal it as something other than the past is typically taken to be in the present, and no such resolute reclamation of the true lineaments of past and present can be enacted except by grasping the future in the light of one's fate as an individual and the destiny of one's community. So, if an historical investigation is to reveal the true heritage of the present, those prosecuting it must themselves embody an authentic mode of human historizing.

Heidegger's idea is that true history allows past, present and future reciprocally to question and illuminate one another, and is thus at once a manifestation of, and a preparation for, anticipatory resoluteness. By doing her job authentically, the historian reveals the past as harbouring the real potential of her present and thus prepares the way for herself and her community to struggle with their destiny. But, since she is herself a historizing (i.e. a temporalizing) being, her selection of an object of historical study will be determined by her orientation to present and future; so her capacity to grasp the particular past possibility which embodies the best destiny of her community, and to disclose it as such, presupposes that she has a resolute grasp of her own present and an anticipatory grasp of her own future:

> Only by historicality which is factual and authentic can the history of what has-been-there, as a resolute fate, be disclosed in such a manner that in repetition the 'force' of the possible gets struck home into one's factical existence – in other words, that it comes towards that existence in its futural character.

(BT, 76: 447)

If historizing and historiology are related in a circle of mutual presupposition, it is always either vicious or virtuous. Either the absence of authentic historizing blocks off the possibility of authentic historiology and is reinforced by so doing, or its presence brings about authentic historiology and thereby reinforces its own reality and wider dissemination. But this circularity suggests a paradox: if authentic historizing presupposes authentic historiology, but only an authentically historizing Dasein can engage in authentic historiology, how can authentic historiology ever get started? The immediate answer is: by the historian shattering herself against death as her ownmost possibility, and thereby being brought to approach the task to which she has dedicated her life with anticipatory resoluteness. She would then understand that her ability to accept her own individual fate cannot be separated from her community accepting its destiny, and that this joint acceptance is made possible only by the successful exercise of the skills that she and her colleagues possess, and the widespread dissemination of the results of their exercise. In other words, what allows Dasein to break into the circle of authentic historiology and authentic historizing is just what allows authenticity to break in upon any human being: the impact of the voice of conscience, the reticent anxiety induced by Dasein's confrontation with the true depths of its own finitude.

But this returns us to the paradox we diagnosed when examining Heidegger's earlier treatment of conscience. If inauthentic Dasein has repressed its capacity for authenticity, how can it utter or hear the call of its conscience, which is the voice of that repressed capacity? My suggested resolution was to modify Heidegger's analysis so as to allow that the voice of conscience might emanate from an external source – from someone else with an interest in her interlocutor's overcoming her inauthenticity and freeing her capacity to live a genuinely individual life, someone prepared to offer herself as an exemplar of what such an authentic mode of existence might be like. At that earlier stage, I had to admit that Heidegger seemed explicitly to reject this modification; but it did dovetail smoothly with much of what he actually said about the voice of conscience.

Now, I think we can say that Heidegger's discussion of historicality and historiology deliberately commits him to just such a resolution of the paradox. For he ends it with a sudden (and, within the precincts of *Being and Time*, unique) cluster of predominantly respectful references to other thinkers. Nietzsche takes the stage as someone whose analysis of the 'use and abuse of historiology for life' contains in embryo the core of Heidegger's own analysis; and, most prominently, the chapter ends with an admiring six-page discussion of Wilhelm Dilthey's and Count Yorck von Wartenburg's conceptions of the human sciences in general and the science of history in particular.

Looked at in itself, the location, structure and content of this concluding discussion is deeply puzzling. First, and assuming for the moment that Heidegger correctly represents the thought of Dilthey and Yorck therein, it adds nothing to the conclusions already established earlier in the chapter; at best it shows only that they were in some very dim and indirect ways presaged in the work of these two men. Second, despite the fact that Heidegger interprets Yorck as merely clarifying the underlying message of Dilthey's work, the quotations Heidegger assembles from Yorck's letters to Dilthey have a continuously critical tone. Third, the discussion focuses upon what seem very marginal texts: instead of examining Dilthey's more famous works, Heidegger's attention is on Yorck – and Yorck's letters at that. And, finally, Heidegger's own voice virtually disappears from these concluding pages; his purported discussion of Dilthey's and Yorck's thought is in fact little more than a sequence of quotations from Yorck.

If, however, we place this discussion in the context of the voice of conscience, these difficulties disappear. What Heidegger is offering is an example of how the voice of conscience can break in upon historiology. Yorck's letters to Dilthey are his attempt to point out for his friend's benefit how he might break free from a broadly inauthentic understanding of historiology and historicality by developing those aspects of his views that are closest to what Yorck sees as the truth of these matters. His critique is thus not coercively and futilely external (which would amount to his failing to respect his friend's autonomy), but calibrated to those aspects of Dilthey's own worldview that have the most potential for positive internal development. And, by presenting himself as disclosing points that are already implicit in Dilthey's own work, as in effect his friend's best interpreter, Yorck shows that his own position is not based upon superior expertise. On the contrary, he implies that he could not have attained the position from which he criticizes his friend without standing on his friend's shoulders. In this sense, the position to which Yorck is attempting to attract Dilthey is nothing more than Dilthey's own best possibility – his unattained but attainable self.[2]

This implies more generally that progress towards authenticity in any part of human existence, including historiology, is essentially historical. Yorck's further progress towards the existential truth about the science of history and human existence is itself produced by critically appropriating possibilities disclosed by the past. His position is the result of repeating the past in a moment of vision about the present that is oriented towards the best destiny of himself qua historian, the discipline of which he is a member, and the culture of which that discipline is such an important component. Putting these points together, the final implication of Yorck's example is that for an historian to be authentic is for him to act as the voice of conscience to the past (and thus to the present) of his discipline and its culture. To work with anticipatory resoluteness as an historian amounts to criticizing the past from the perspective of its own best possibilities with a view to galvanizing the present from the perspective of its destined future. And Yorck's example thereby confirms that genuine repetition of the past is no mere reiteration of it. Precisely because the situation of the historian differs from that of those inhabiting the past world he strives to understand, his grasp of the past could never simply coincide with theirs; but it remains nonetheless an understanding of what they understood (since it reveals a possibility inherent in it).

But, of course, this example of the voice of conscience in historiology and of an historian's authentic enactment of his historicality is one that Heidegger provides for his readers; and he does so by presenting Yorck's own position as an unresolved precursor of his own insights. In other words, by placing his account of Dilthey and Yorck at the end of his own investigation of historiology and historicality, he places Yorck in exactly the position that Yorck himself placed Dilthey. Heidegger offers an implicit critique of Yorck, but one which presents itself as internal, devoted to developing Yorck's own best possibilities, and so as one to which Heidegger himself could not have attained without Yorck's own work and example. He thus offers himself as the voice of conscience to Yorck, as an example of authentic historiology (someone capable of renewing the discipline of history by recovering the most fruitful of its past possibilities, even from such unpromisingly marginal documents as private correspondence, and projecting it into the future), and as attempting thereby to befriend his culture – to tear it away from its present forgetfulness of its past and to awaken it to its destiny. But, in so doing, Heidegger implicitly acknowledges that his own best insights into historiology and historicality did not spring fully formed from his own intellect. He presents Dilthey and Yorck as the voice of conscience that awakened him from inauthenticity, and thus bolsters his implicit claim to be the authentic voice of conscience to his readers by implicitly denying that he occupies any position of personal superiority or expertise. He thus avoids suggesting that his readers are somehow in an inferior position to his own, a suggestion which seemed to be encoded into his earlier discussion of the voice of conscience and which implied that he was not sufficiently respectful of the autonomy of those he was addressing and claiming to befriend. We can therefore conclude that the modifications to the model of the voice of conscience which we offered earlier were simply an anticipation of Heidegger's own self-criticism. Even the author of *Being and Time* is not capable of escaping inauthenticity entirely by his own efforts.

However, when I introduced the idea of the friend to solve the problem of bootstrapping inauthentic Dasein into authenticity, I noted that it appears simply to displace the problem it attempts to solve on to the friend. For, if inauthentic Dasein's transformation to authenticity presupposes a friend, how did that friend attain authenticity? Heidegger's discussion of Dilthey and Yorck suggests the following answer: through the intervention of another friend – Yorck can befriend Heidegger because he was befriended by Dilthey. But such chains of friendship must surely have a beginning, a first link; and a first friend would necessarily be an unbefriended friend, someone who managed the transformation into authenticity unaided. But it was the impossibility of such a self-overcoming of self-imposed lostness that caused our problem in the first place.

This worry is misplaced. A first or self-befriending friend would be required only in a world in which human inauthenticity was universal and absolute; and Heidegger's conception of human existence neither entails nor permits such a possibility. He does claim that lostness in the they-self is Dasein's typical position, even that it inherently tends towards fallenness, because its social roles are essentially impersonal; but this makes authenticity a rare and fragile achievement, not an impossible one. And no community of beings to whom an understanding of their own Being necessarily belongs could utterly lose a sense of themselves as capable of authenticity. Whether in disregarded texts, moribund institutions or marginalized individuals (like Dilthey and Yorck), some vestiges of that self-interpretation will survive for as long as human beings do, and thereby make it possible for chains of friendship to maintain and develop themselves. The friendship model of conscience does not therefore require the self-defeating invocation of a self-befriending friend; the human world could never be entirely incapable of disrupting the inveterate repressions of inauthenticity.

ON BEING WITHIN TIME (§§78–82)

In his final chapter, Heidegger concludes his analysis by relating his existential understanding of time to that which prevails not just in Dasein's ordinary life but in disciplines devoted to theorizing about the fundamental structures of that life (e.g. philosophy). In everyday life, for example, we talk of entities as something we encounter in time, and describe our own activities in ways which imply that time is something we can possess or lose – as when we say that we have no time to do something, or that doing something will take a certain amount of time. These formulations suggest a conception of time as something objective – either a medium in which things are immersed or a substance or property that we can grasp, take or lose. This conflicts with the existential conception of temporality as the ontological foundation of Dasein's Being as care. In addition, prevailing philosophical conceptions of time (on Heidegger's view, still rooted in the work of Aristotle) portray it as a sequence of self-contained units, a series of 'nows' that emerge from the future, present themselves to the individual and disappear into the past. This flatly contradicts the existential conception of temporality as an articulated ecstatic unity. If, however, all modes of human existence are grounded in temporality, then the lives of those who adopt an average everyday conception of time, as well as the interpretative structures presupposed by its theoretical thematization and development, must be modes of temporalizing – however inauthentic. But how is it possible for beings, whose relation to time is of the sort Heidegger has been claiming, to misunderstand the nature of their own existence in just these ways? How might such misunderstandings have developed, and how can their existential realization be understood in terms of temporality?

Our everyday understanding of time is manifest in the way we locate events and other phenomena in temporal terms: we talk of things happening now, of something that has not yet happened but is to happen then, and of things that happened previously or on a former occasion. Clearly these three broad types of reference to time form a single interrelated framework – what Heidegger calls 'datability': what is awaited or expected to happen (at a certain time) does indeed happen, and thereafter can be referred to as something that happened on that former occasion. But the datability of events is at least implicitly founded upon the present moment, the 'now': the 'then' is understood to be the 'not-yet now', and the 'on that former occasion' is a reference to the 'no-longer now'. This is because, in everyday life, Dasein is typically concerned with the entities among which it finds itself, and with the task for which they are ready-to-hand or unhandy; so it is naturally primarily oriented towards that with which it is presently concerned, with future and past events primarily regarded as phenomena which either will be or were the focus of its present concern.

Datability does not, however, immediately imply an exclusive focus upon time as comprising a succession of moments or instants; for tasks occupy periods of time as much as they do moments. When we talk of having no time to do something, or of having lost track of the time while doing something, we articulate a sense of time as something that spans moments, something which endures or lasts. Moreover, what 'now' means will often vary according to our current preoccupations – 'now' might pick out the instantaneity of a match being struck or the hours occupied by dining at a restaurant. And the datability and spanning of time is essentially public. When we talk of something's having come to pass 'now', the time we thereby pick out is equally accessible to others: the beginning of the Second World War, the time at which the dinner party moved on to dessert, the time it took for someone to repair her roof – these are not private or inherently subjective matters but issues of public dispute and agreement. It is this which most firmly grounds our everyday sense of time as something objective or autonomous – a frame of reference to which we adjust ourselves rather than one we impose upon our experience.

These three elements of the everyday conception of time are thus tightly interwoven; and at least the first two can be interpreted as rooted in temporality. The very fact that the three dimensions of datability are inherently interrelated reflects the interarticulation of the three temporal ecstases; while the notion that time is periodic or spanned manifests the fact that Dasein's existence is a matter of its stretching along and being stretched along its days. Pointing to a structural analogy between the two conceptions, however, does not amount to providing a derivation of the former from the latter – a proof that only an existential understanding of time as temporality can account for the everyday conception of time. And what of its inherently public nature? How does the possibility of our orienting ourselves by reference to such datable spans of time, our seeming ability to come across time in our dealings with the world, relate to the temporalizing roots of Dasein's Being? Heidegger's answer utilizes the inherent worldliness of human existence to develop a highly speculative, but peculiarly powerful, brief history of the development of Dasein's reckonings with time – what one might call an enabling myth of chronology.

According to that myth, Dasein's most primitive mode of reckoning with time is astronomical; and this is because its Being is care. Always already thrown into the world, and typically lost in a kind of fascinated absorption with the entities it encounters there, human beings relate to those entities in terms of their possible and actual involvement with their own tasks or projects. But they can hardly engage in practical activity if they cannot perceive their world of work. They must therefore reckon with periods of darkness and light, awaiting the passage of night and the arrival of the dawn; and this means reckoning with dawn and dusk as the time to begin work and to put it aside:

> Dasein dates the time which it must take, and dates it in terms of something it encounters within the world . . . as having a distinctive involvement for its circumspective potentiality-for-Being-in-the-world. Concern makes use of the Being-ready-to-hand of the sun which sheds forth light and warmth. The sun dates the time which is interpreted in concern. In terms of this dating arises the most 'natural' measure of time – the day.

> (BT, 80: 465)

The time-cycle reckoned with in everydayness is thus essentially daily or diurnal – the cycle of days and of months, as well as the day's internal divisions, are measured in accordance with the sun's journeying across the heavens. Thus, the diurnality of everyday Dasein embodies a definite kind of periodicity or spanning. And, since the basis of this time-reckoning is astronomical, it is inherently public: the rising, progress and setting of the sun are not exclusive to any particular individual or world of equipment. In effect, then, the sun is Dasein's first and most fundamental clock; but this mode of reckoning with time as public, spanned and datable has an obvious relation to Dasein's projects. The position of the sun is to be reckoned with because given degrees of its brightness and warmth are variously appropriate to a given task; early summer mornings are best for harvesting, but a winter dusk is perfectly suited to feeding cattle. Thus, reckoning with the sun presupposes the network of 'in-order-to' and 'for-the-sake-of' relations which make up the interpersonal structures of significance, grounding all of Dasein's practical activities – the worldhood of the world. In other words, the time with which Dasein is reckoning is inherently worldly – it is world-time.

So the first clock becomes accessible only because human existence itself is inherently worldly, inherently a matter of encountering entities; the sun is a clock that is always disclosed to Dasein as a ready-to-hand part of Nature and the common social environment. And human worldliness is founded upon the care-structure, which is itself founded upon temporalizing temporality. In short, the accessibility of a clock is not the precondition for time; human temporality is the precondition for any and every form of clock-time.

In Heidegger's myth, all future developments of clock-time – the use of shadows cast by the sun, sundials, clocks and pocket watches, digital and atomic clocks – build upon the datability, spannedness and publicity established by the first uses of the sun as a clock. Even methods of time-measurement that make no explicit reference to the sun necessarily draw upon knowledge of the processes of the natural world which is first illuminated by, and disclosed simultaneously with, this natural clock. The inherently public nature of everyday time is thereby reinforced; but this is achieved not by detaching clock-time from its worldliness, but by relying upon that connection. Reckoning with electrical impulses or the decay of atomic nuclei is no less dependent upon the human being's disclosedness of its world, and the time thus measured is accordingly no less world-time. And since such modes of reckoning presuppose time's inherent worldliness, they presuppose the essentially temporal foundation of human existence as Being-in-the-world.

This means that both the theorizing and the forms of life that presuppose the everyday conception of time (however technically advanced the modes of time-reckoning they involve) are enactments of a specific form of Dasein's threefold ecstatic temporality. But, if every mode of the care-structure is either authentic or inauthentic, the same must be true of this mode of temporalizing. And, according to Heidegger, it is deeply inauthentic – a reflection of Dasein's lostness in the 'they'. The mode of datability involved is spanned and public; but its publicity is understood as something entirely objective – something to be met with in the world, something human beings must confront and which has no relation to their own existential foundations. Similarly, its being spanned is understood primarily in relation to the period of time required for the completion of a task rather than as something which most basically relates to Dasein's existence as stretched along the sequence of its days; time's periodicity is thus detached from the fundamental question of Dasein's challenge to establish and maintain self-constancy. And both ways of levelling-off or repressing the true significance of time as temporality derive from the basic form of everyday time's datability – the priority it gives to the 'now'.

As we saw earlier, the 'then' and the 'on a former occasion' are understood in terms of the now – the former as a 'not-yet now' and the latter as a 'no-longer now'. That amounts to emphasizing the temporal ecstasis of the present, and enacting that ecstasis in the form of making-present – something that goes together with a forgetting of the past and an awaiting of the future. People caught up in this mode of datability are completely absorbed in the present object of their concern, and so entirely dismiss that which is no longer present (since it can be of no use to this concern), while comprehending what is to come entirely in terms of its usefulness for their present concern. The significance of the future and the (in)significance of the past are thus determined solely by what is presently preoccupying them: the past becomes instantly obsolete, and the future more and more eagerly (but more and more unquestioningly) leapt upon as grist to contemporary mills. The result is an effective dispersal or dissolution of the self's individuality in the publicly dictated demands of the task with which it is fascinated:

> The irresoluteness of inauthentic existence temporalizes itself in the mode of a making-present which does not await but forgets. He who is irresolute understands himself in terms of those very closest events and be-fallings which he encounters in such a making-present, and which thrust themselves upon him in various ways. Busily losing *himself* in the object of his concern, he *loses his time* in it too.
>
> (BT, 79: 463)

What is missing here is any possibility of relating to the present in and as a moment of vision – a grasp of its resources as a context for existentiell choice, the scene for a penetrating repetition of the past that might liberate real but hidden possibilities for the future.

Someone adopting this mode of temporalizing, someone gripped by anticipatory resoluteness, breaks through the levelling-off of temporality as time and thereby tears herself away from lostness in the 'they', re-establishing self-constancy by having time for what the situation demands and having it constantly. But the individual who is absorbed by and enacts the everyday conception of time is entirely closed off from any such understanding of time and of her own relation to it – and so from any possibility of wrenching herself towards an enactment of it. Living in accord with the datability, spannedness and publicness of everyday time is a mode of temporalizing that represses any possibility of understanding itself as such.

Accordingly, when the task of thematizing an understanding of time emerges and is addressed in such disciplines as philosophy, it is done in such a way that even the basic structure of everyday time is overlooked. For any would-be philosopher of time naturally abstracts her conception of her topic from those modes of time-reckoning with which she is most familiar – from circumspective, concernful clock-using. And since these clocks are typically non-natural or non-solar, what appears central to our telling the time is our making-present a moving pointer – following the sequence of positions that a pointer moves through on a dial. But, when one follows such a pointer, one checks off a successive series of 'nows': one would say 'Now it's here, now here' and so on. And thus emerges a conception of time as a successive flow of self-contained and present-at-hand 'nows'. It is not built into our unthematized reckonings with time in the public work-world, but developments within that world designed to make time-reckoning more ready-to-hand (i.e. the development of clocks) make it all but unavoidable when we thematize time as such. When we do so, not only the idea of clock-time as grounded in temporalizing, but also that of time as public, spanned datability, is repressed. For the datability of time presupposes the interrelatedness of its three dimensions and their involvement with structures of significance (i.e. 'then' means both 'not-yet now' and 'then, when I tried to'); but no sequence of atomized instants could manifest such interrelatedness or such significance.

Thus, in the philosophical tradition, even an accurate understanding of everyday time – let alone a properly existential conception of time as temporality – is covered over. Heidegger offers Aristotle's and Hegel's analyses of time and the human relation to time as paradigms of such repression. This is symptomatic of Dasein's more general tendency to misunderstand its own Being – a tendency deriving from the nature of Dasein's Being as care. For Dasein tends to interpret everything it attempts to thematize in the terms appropriate to that with which it is most familiar – that is, in terms of readiness-to-hand and presence-at-hand. And, just as the readiness-to-hand of entities is mistakenly interpreted by average everyday Dasein in terms of presence-at-hand, so the same fate befalls time:

> Thus the 'nows' are in a certain manner *co-present-at-hand*; that is, entities are encountered, and so too is the 'now'. Although it is not said explicitly that the 'nows' are present-at-hand in the same way as Things, they still get 'seen' ontologically within the horizon of the idea of presence-at-hand.

> (BT, 81: 475)

On this understanding of time, of course, there are only two ways of conceiving its ontological status. Either it is objective in the way that material objects are, or it is subjective in the way that psychical experiences are; it is present-at-hand in the world, or it is present-at-hand in the subject. Whereas, for Heidegger, time is both objective and subjective – but not at all in the way philosophers envisage it. It is objective in the sense that it is inherently worldly: world-time is more objective than anything we might come across within the world because it is the ecstatico-horizonal condition for the possibility of coming across entities in the world. And it is subjective in the sense that the ontological roots of its worldliness lie in the human way of being: it is more subjective than anything in the psychic life of an individual because it is the condition for the possibility of the existence of any being whose Being is care.

On this account, there is a clear sense in which both Dasein and the entities it encounters are in time (since entities are datable in their comings and goings, and Dasein is stretched along temporally); and there is an equally clear sense in which they are not (since the datability of entities is ontologically derived from the temporality of Dasein's Being, while the temporality of Dasein's Being means that Dasein is [or exists as] time rather than existing in time). In other words, only an account of the existential foundations of time as temporality grasps the underlying structure of world-time in a way that avoids the Scylla of vicious reification and the Charybdis of subjectivist volatilization. Only an account of the human way of being as temporality can explain the sense in which human beings and the entities they encounter are (and are not) within time.

NOTES

1   Cf. D. Parfit, *Reasons and Persons* (Oxford: Clarendon Press, 1984).
2   This marks another point at which my implicit broad reliance upon Cavell's model of perfectionism brings me to the point of finding his own terminology ready-to-hand for my purposes; see the references cited in Chapter 4, note 4.

8

CONCLUSION TO
DIVISION TWO:
PHILOSOPHICAL ENDINGS
– THE HORIZON OF
BEING AND TIME

(*Being and Time*, §83)

HUMAN BEING AND THE QUESTION OF BEING IN GENERAL

Heidegger concludes his phenomenological investigation of the human way of being by making it absolutely clear that his uncovering of temporality as its basis is both an end and a beginning. It is an end in that it provides the most fundamental understanding that he has been able to develop of the nature of human existence. Over five hundred closely argued pages, he has argued that Dasein is essentially worldly, that this worldliness is founded upon the tripartite care-structure, and that this care-structure is itself founded upon the threefold ecstatic temporalizing of temporality. But this analysis of Dasein's conditionedness or finitude was never an end in itself. It was, rather, his way of addressing the broader and more fundamental question of the meaning of Being in general; and *Being and Time* ends by re-posing that question.

Heidegger offered three reasons for regarding an existential analytic of human being as a way of working out the question of the meaning of Being in general. Human beings can encounter other entities in their Being and are fated to confront their own Being as an issue, so they are doubly related to Being in everything that they do; and, since any investigation of the meaning of Being is itself a possible mode of human existence, a proper understanding of its limits and potentialities requires a prior grasp of the nature of human existence as such. This ontico-ontological priority of Dasein, as Heidegger calls it, means that an investigation of human existence is not just a convenient starting point from which to address the question of the meaning of Being in general – it is indispensable.

By the very same token, however, even a provisional answer to the question of the meaning of the Being of Dasein cannot in itself amount to an answer to the question of the meaning of Being in general. The two questions are internally related, but not identical. The latter asks for an account of the underlying differentiated unity of whatever it is that is made manifest through the manifestation of any and every being in its Being – not just that of the being whose Being is Dasein. Nevertheless, since human beings can grasp any and every entity in its Being, understanding the ontological grounds of that capacity might at least equip us to pose the question of the meaning of Being in a fruitful manner. In this sense, the existential analytic of Dasein puts us on the way to answering the question with which Heidegger is primarily concerned. And, of course, the critical term required for posing this question fruitfully turns out to be that of time – or, rather, temporality:

> Something like 'Being' has been disclosed in the understanding-of-Being which belongs to existent Dasein as a way in which it understands. Being has been disclosed in a preliminary way, though non-conceptually; and this makes it possible for Dasein as existent Being-in-the-world to comport itself *towards entities* – towards those which it encounters within-the-world as well as towards itself as

existent. *How is this disclosive understanding of Being at all possible for Dasein?* Can this question be answered by going back to the *primordial constitution-of-Being* of that Dasein by which Being is understood? The existential-ontological constitution of Dasein's totality is grounded in temporality. Hence the ecstatical projection of Being must be made possible by some primordial way in which ecstatical temporality temporalizes. How is this mode of the temporalizing of temporality to be Interpreted? Is there a way which leads from primordial *time* to the meaning of *Being*? Does *time* itself manifest itself as the horizon of *Being*?

(BT, 83: 488)

When thematized, Dasein's understanding of Being, its openness to its world, is shown to depend upon the care-structure, which is in turn grounded in ecstatic temporality. The horizonal structure of the world (the inexhaustible, self-concealing clearing within which Being is manifest as the Being of some entity or other) is grounded in the horizonal structure of temporality (Dasein's endless standing-outside itself in the three interlinked temporal schemas); temporality is the fundamental condition for the possibility of grasping beings in their Being. Heidegger is not here identifying Being and time. His book has shown that temporality is the ground of Dasein's understanding of beings in their Being, and an understanding of beings in their Being is not the same as an understanding of Being – any more than an understanding of Being is Being itself. Nevertheless, Being and time cannot be entirely distinct, because the concept of Being and the concept of an understanding of Being as manifest in beings are internally related. Being itself can never be encountered except as the Being of some being or other; and, in so far as any attempt to answer the question of the meaning of Being will be the act of some particular human being, it must articulate *an understanding* of the meaning of Being. Accordingly, Heidegger ends his book by asking the question of the meaning of Being in the form that his existential analytic of Dasein suggests – by asking whether time *manifests* itself as the *horizon* of Being.

To find that this complex, dense and difficult text ends with the posing of the very question with which it began, rather than with any attempt to answer it, may seem a profoundly unrewarding conclusion for its readers. But the book as a whole has provided a great deal of information about the human mode of being on the way to re-posing this question; and some of that information made it inevitable that *Being and Time* would end in exactly this way. To begin with, author and reader have been collaborating in an ontological investigation – developing a particular interpretation of Being as it manifests itself in and through Dasein; and, according to that interpretation, interpretations generally move within a hermeneutic circle or spiral. But this means not just that there can be no interpretation-free point at which to commence the hermeneutic task, but also that there can be no definitive end to it. Any text, action or practice under interpretation forms part of a complex network of objects and activities that is in turn founded upon structures of significance which are not reducible to a finite list of elements or rules; so each step forward in the interpretative enterprise inevitably opens up new vistas of meaning that call for further exploration. In this sense, interpretation is essentially horizonal, and so in principle incapable of attaining absolute completion. Indeed, if interpretation can never be absolutely terminated, the fact that a text ends by posing further questions does not show that it is essentially incomplete. For, if there can be no conclusions that do not raise further questions, then an interpretative text's final posing of a question cannot show that it has not reached a conclusion, or been brought to a perfectly adequate terminus. Accordingly, for Heidegger to end in any other way than by pointing out the new vistas of meaning that his interpretation of the Being of Dasein has opened up, would be for the form of his text to contradict, and so to indict, its content.

Even if we acknowledge this, however, we might think that the task of exploring the new vistas that are visible from this textual terminus is primarily Heidegger's; and we might then be tempted to search out the other texts that Heidegger authored in which (scholars claim, and not wrongly) he does just that. As I mentioned in the Introduction, there are a number of texts from the late 1920s that might justifiably be regarded as providing the essential elements of the four further divisions that are mentioned in Heidegger's opening delineation of his project, but are absent from *Being and*

Time itself. But there are also texts from the same period – perhaps most obviously his inaugural lecture at Freiburg, entitled *What is Metaphysics?* – which explicitly take up and elaborate a connection that we have seen to be implicit in, and deeply determinative of, the course of *Being and Time* itself: for, if Dasein is the Being for whom Being is an issue, and if there is an uncanny intimacy between the Being of Dasein and nullity, negation and nothingness, then there must be a deep affinity between Being and 'the nothing'.

As we saw most explicitly in Chapter 5, however, Heidegger's realization of the internal relation between Dasein and nothingness was also a realization that this relation placed the very possibility of a phenomenological analysis of the Being of Dasein in question. For nothingness is neither a phenomenon nor of the logos – neither an entity that might appear to us as it is in itself, nor the object of a possible discursive act. Heidegger's response to this problem in *Being and Time* is to attempt to represent the nothing as the beyond of phenomenological representation – as the unrepresentable condition for the possibility of Dasein's comprehending and questioning grasp of beings in their Being. He aims to achieve this goal by presenting Division Two as pointing towards that which lies beyond Division One: it neither identifies some specific feature(s) of Dasein's Being omitted by Division One, nor merely reiterates Division One's conclusions about Dasein's Being in a more ontologically penetrating manner, but, rather, repeatedly brings us up against the unrepresentable horizon of every element of the analysis in Division One. In this respect, Division Two does not simply illustrate the hermeneutic insight that, no matter how much we say about Dasein's Being, there is always more to be said; it, rather, enacts the thought that there is something inherently enigmatic about the Being of Dasein – something necessarily beyond the grasp of that being itself, and hence necessarily beyond the grasp of any existential analytic of its Being.

One might say that, for Heidegger, any adequate account of Dasein's Being must embody a continuous or pervasive acknowledgement of its ineluctable inadequacy: hence the uncanny non-coincidence of Division Two with Division One; hence his blatantly self-subversive talk in Division Two of impossible possibilities, of unrepayable debts and silent voices, of repetition without reiteration; hence his emphasis on Dasein's self-transcendence, its non-self-identity, its inability to coincide with itself, its essentially ecstatic unity. But such a sense of Dasein's Being as inherently enigmatic would not encourage the thought that further turns around the spiral of understanding initiated in Division One might bring us to an ever-deepening grasp of that Being. It would rather suggest the need to be sure that what phenomenological analysis discloses as enigmatic really is enigmatic and not just indicative of the representational limitations of phenomenology. And that would mean devoting more explicit reflection to the means of representation at Dasein's disposal – perhaps by paying closer attention to the nature of language, perhaps by looking at the variety of modes of human linguistic and non-linguistic communication, perhaps by fashioning a variety of alternative modes of philosophical discourse in order to discover whether each is fated to subvert itself in the manner of phenomenology when it attempts to probe (what phenomenology calls) the Being of Dasein and hence Being as such. Those familiar with Heidegger's writings after the supposed 'turn' in his thought might recognize each of these possibilities as actualized in that vast array of texts.

There is one further moral that might be drawn from *Being and Time*'s open-ended ending. To appreciate it, we must recall his discussions of what might constitute human authenticity, apply their conclusions to ourselves as human beings presently engaged in the task of reading philosophy, and also recall that the words ordered to form the text we are reading implicitly claim to be articulations of the voice of philosophy's conscience. Then we might interpret its author not as posing a question to which he intends to provide a concrete answer elsewhere, in some other arrangement of words at some other time and place, but as posing a question which he expects us to answer. After all, a question is typically posed because the questioner would like the hearer to supply an answer; by no means all questions are rhetorical, or otherwise posed solely in order that the *questioner* may provide the answer. And, as Heidegger understands his role as the voice of conscience in philosophy, his most important responsibility is to restore the autonomy of his readers, to wrest them away from an unquestioning reliance upon the deliverances of the tradition and their colleagues. He would hardly live up to that responsibility if he merely substitutes a reliance upon him for their previous reliance upon others. In other words, an important part of his reason for concluding *Being and Time* with a question might well be that it constitutes a rebuke to its readers, a way of warning his would-be followers against relying upon him to provide all the answers they seek in their philosophical investigations – without realizing that such a reliance upon others is an abdication of self-responsibility as a thinker, a refusal of the very insight about self-reliance that they claim to have acquired. In short, the constituent terms of Heidegger's concluding question indicate the way to go on from his words; but the fact that they constitute a question indicates that it is a route we should be prepared to trace out for ourselves. In this sense, the conclusion of *Being and Time* demonstrates that the path of true thinking is one that each reader must take for herself.

BIBLIOGRAPHY

BOOKS BY HEIDEGGER REFERRED TO IN THE TEXT

Being and Time, trans. J. Macquarrie and E. Robinson (Oxford: Basil Blackwell, 1962).

The Basic Problems of Phenomenology, trans. A. Hofstadter (Bloomington, Ind.: Indiana University Press, 1982).

Kant and the Problem of Metaphysics, trans. R. Taft (Bloomington, Ind.: Indiana University Press, 1990).

COMMENTARIES ON *BEING AND TIME* (AND OTHER HEIDEGGER TEXTS)

Dreyfus, H., *Being-in-the-World* (Cambridge, Mass.: MIT Press, 1991).

Philipse, H., *Heidegger's Philosophy of Being* (Princeton, N.J.: Princeton University Press, 1998).

Poggeler, O., *Martin Heidegger's Path of Thinking*, trans. D. Magurshak and S. Barber (Atlantic Highlands, N.J.: Humanities Press International, 1987).

Polt, R., *Heidegger: An Introduction* (London: UCL Press, 1999).

Richardson, J., *Existential Epistemology* (Oxford: Clarendon Press, 1986).

Steiner, G., *Heidegger* (London: Fontana, 1978; revised edition, 1994).

COLLECTIONS OF ARTICLES ON HEIDEGGER

Dreyfus, H. and Hall, H. (eds), *Heidegger: A Critical Reader* (Oxford: Blackwell, 1992).

——, and Wrathall, M. (eds), *The Blackwell Companion to Heidegger* (Oxford: Blackwell, 2005).

Guignon, C., *The Cambridge Companion to Heidegger* (Cambridge: Cambridge University Press, 1993).

Sallis, J., *Reading Heidegger: Commemorations* (Bloomington, Ind.: Indiana University Press, 1994).

OTHER BOOKS REFERRED TO IN THE TEXT

Cavell, S., *Conditions Handsome and Unhandsome* (Chicago, Ill: Chicago University Press, 1990).

Golding, W., *The Spire* (London: Faber and Faber, 1964).

Honderich, T. (ed.), *Morality and Objectivity: Essays in Honour of J. L. Mackie* (London: Routledge, 1985).

Kant, I., *Critique of Pure Reason*, trans. N. Kemp Smith (London: Macmillan, 1929).

Kierkegaard, S., *Concluding Unscientific Postscript*, trans. H. V. and E. H. Hong (Princeton, N.J.: Princeton University Press, 1992).

Mulhall, S., *Faith and Reason* (London: Duckworth, 1994).

Parfit, D., *Reasons and Persons* (Oxford: Clarendon Press, 1984).

Ryle, G., *The Concept of Mind* (London: Hutchinson, 1949).

Strawson, P. F., *Individuals* (London: Routledge and Kegan Paul, 1959).

Taylor, C., *Philosophical Papers Vols I and II* (Cambridge: Cambridge University Press, 1985).

——, *Sources of the Self* (Cambridge: Cambridge University Press, 1989).

Weston, M., *Kierkegaard and Modern Continental Philosophy* (London: Routledge, 1994).

Wittgenstein, L., *Tractatus Logico-Philosophicus*, trans. C. K. Ogden (London: Routledge and Kegan Paul, 1922).

——, *Philosophical Investigations*, trans. G. E. M. Anscombe (Oxford: Basil Blackwell, 1953).

INDEX

Aristotle in *Routledge Guides to the Great Books*

The Routledge Guidebook to Aristotle's Nicomachean Ethics

Gerard J. Hughes, University of Oxford

Written by one of the most important founding figures of Western philosophy, Aristotle's *Nicomachean Ethics* represents a critical point in the study of ethics which has influenced the direction of modern philosophy. The *Routledge Guidebook to Aristotle's Nicomachean Ethics* introduces the major themes in Aristotle's great book and acts as a companion for reading this key work, examining:

- The context of Aristotle's work and the background to his writing
- Each separate part of the text in relation to its goals, meanings and impact
- The reception the book received when first seen by the world
- The relevance of Aristotle's work to modern philosophy, its legacy and influence.

With further reading included throughout, this text is essential reading for all students of philosophy, and all those wishing to get to grips with this classic work.

February 2013 – 336 pages
Pb: 978-0-415-66385-4| Hb: 978-0-415-66384-7

The Routledge Guidebook to Hegel's Phenomenology of Spirit

Robert Stern, University of Sheffield

The *Phenomenology of Spirit* is arguably Hegel's most influential and important work, and is considered to be essential in understanding Hegel's philosophical system and his contribution to western philosophy. The *Routledge Guidebook to Hegel's Phenomenology of Spirit* introduces the major themes in Hegel's great book and aids the reader in understanding this key work, examining:

- The context of Hegel's thought and the background to his writing
- Each separate part of the text in relation to its goals, meaning and significance
- The reception the book has received since its publication
- The relevance of Hegel's ideas to modern philosophy

With a helpful introductory overview of the text, end of chapter summaries and further reading included throughout, this text is essential reading for all students of philosophy, and all those wishing to get to grips with Hegel's contribution to our intellectual world.

February 2013 – 276 pages
Pb: 978-0-415-66446-2| Hb: 978-0-415-66445-5

Available from all good bookshops

Plato in *Routledge Guides to the Great Books*

The Routledge Guidebook to Plato's Republic

Nickolas Pappas, University of New York

Plato, often cited as a founding father of Western philosophy, set out ideas in the *Republic* regarding the nature of justice, order, and the character of the just individual, that endure into the modern day. *The Routledge Guidebook to Plato's Republic* introduces the major themes in Plato's great book and acts as a companion for reading the work, examining:

- The context of Plato's work and the background to his writing
- Each separate part of the text in relation to its goals, meanings and impact
- The reception the book received when first seen by the world
- The relevance of Plato's work to modern philosophy, its legacy and influence.

With further reading included throughout, this text follows Plato's original work closely, making it essential reading for all students of philosophy, and all those wishing to get to grips with this classic work.

February 2013 – 320 pages
Pb: 978-0-415-66801-9| Hb: 978-0-415-66800-2

Available from all good bookshops

Printed in Great Britain
by Amazon